AF412817

PLASMA PHYSICS

GRADUATE STUDENT SERIES IN PHYSICS

General Editor: Professor Douglas F. Brewer M.A., D.Phil.
Professor of Experimental Physics, University of Sussex

PLASMA PHYSICS

E. W. LAING
M.A., Ph. D.

Department of Natural Philosophy
University of Glasgow

SUSSEX UNIVERSITY PRESS

1976

Published for
Sussex University Press
by
Chatto & Windus Ltd
40 William IV Street
London WC2N 4DF

★

Clark, Irwin & Co. Ltd
Toronto

*

Crane, Russak & Co. Inc
347 Madison Ave,
New York, N.Y. 10017
Library of Congress Catalog No. 75-34579
ISBN   0-8448-0852-0

Paperback ISBN 0 85621 057 9

Printed in Great Britain by
REDWOOD BURN LIMITED
Trowbridge & Esher

CONTENTS

PREFACE

During the last few years, I have given courses on plasma physics
to senior undergraduate classes and to first year graduate students,
and it is partly from these courses, and partly from my own interests
in thermonuclear fusion research, that this introductory text has
evolved.   While there are many excellent books on plasma physics
for research workers in the field, they are mostly too advanced or
too detailed to suit the needs of students meeting the subject for
the first time, and it is with these students expressly in mind that
this book has been written.   Consequently, in restricting the
material to suit this purpose, it is neither possible nor desirable
to cover all the many varied topics currently in high fashion, such
as non-linear phenomena and turbulence.   No attempt is made either
to describe even the most basic experimental techniques, though a
research student involved in a plasma experiment will soon realise
the part played by diagnostics in understanding what goes on in his
plasma.

The book itself is essentially in two parts, with the first four
chapters laying the foundations of the subject.   In Chapter 1,
elementary concepts are introduced which will reappear in various
guises throughout.   This is followed by three chapters which
emphasize different facets of plasma behaviour.   Chapter 2 outlines
the subtleties of the motion of a charged particle in given
inhomogeneous electric and magnetic fields.   The fluid behaviour
of a plasma in its interaction with magnetic fields is discussed
in Chapter 3, while Chapter 4, on plasma kinetic theory, introduces
the reader to collective phenomena, in which the particle and fluid
behaviour are intimately related.

The last four short chapters of the book are meant to do no more
than give a glimpse of research applications.   Chapter 5 gives a
brief account of magnetic confinement of plasma in laboratory
experiments;   Chapter 6 touches on several topics in astrophysics;
Chapter 7 outlines the basic ideas involved in laser compression
of plasma to ultra-high densities;   the book ends with a survey of
computational plasma physics.

ACKNOWLEDGMENT

I gratefully acknowledge permission granted by The Institute
of Physics to use, with minor modifications, the article on
Computational Plasma Physics in <u>Plasma Physics</u>, Conference Series
No. 20, as the last chapter of this book.

Chapter 1

AN INTRODUCTION TO PLASMA PHYSICS

1-1 THE IMPORTANCE OF PLASMA AS A STATE OF MATTER

A plasma is any state of matter which contains enough free, charged
particles for its dynamical behaviour on some scale to be dominated by electromagnetic forces.   We shall restrict our attention
to the high temperature, low density, ionised gas in which random
atomic collisions are so energetic that the gas becomes fully
ionised and no consideration of atomic recombination processes is
required.

As examples of naturally occurring plasmas we look mainly to
astrophysics.   The sun and stars are hot enough to be completely
ionised.   Interstellar gas, though normally only lightly ionised
by interstellar radiation, behaves as a medium of high electrical
conductivity.   In fact, most of the universe is in the plasma
state, our planet being in the exceptional category.

Looking nearer home, several examples of geophysical plasmas
occur such as (i) transient plasma caused by lightning; (ii) the
ionosphere; (iii) further out, the Van Allen belts, consisting of
electrons and protons trapped by the earth's magnetic field;
(iv) the solar wind, an emission of charged particles from the sun.

One of the first applications of plasma physics of major
technological importance was in short-wave radio-communication,
though initially the facts were not fully understood.   Why did
radio waves propagate round the earth, when it was expected that
they should propagate in a straight line, and so at best
tangentially to the earth's surface?   The explanation was given in
terms of the ionosphere, and its properties of reflecting electromagnetic waves.   Detailed explanation was given in the magneto-
ionic theory of Appleton and Hartree[1] in 1928.   At the same time,
Langmuir[2] first observed very high frequency electrical
oscillations, plasma oscillations, in gas discharges.

Gas discharges were in fact the earliest artificial plasmas,
though the standard fluorescent light has such a low degree of

ionisation that it is barely a plasma. It is just because of its
high rate of recombination that it is a useful lighting device.
More recently, during the last two decades, plasma physics research
has had a tremendous boost, since it was realised that a reactor
based on controlled thermonuclear fusion had virtually limitless
possibilities as a source of power.

In essence, the thermonuclear reactor depends on the fusion of
deuterium and tritium nuclei into nuclei of higher mass number.
For example, in the case of deuterons, there are two primary
nuclear reactions:

$$d + d \rightarrow t + p + 4{\cdot}04 \text{ MeV}$$

$$\text{and} \qquad d + d \rightarrow He^3 + n + 3{\cdot}27 \text{ MeV},$$

which occur with equal probability in the random collisions in a
deuterium plasma. The d, t reaction is even more favourable:

$$d + t \rightarrow He^4 + n + 17{\cdot}6 \text{ MeV}.$$

However, great technological problems arise. Even assuming that
the plasma is fully ionised, two nuclei will repel each other
electrostatically, so that nuclei will usually collide elastically,
and fusion reactions will not occur in sufficient numbers until the
average kinetic energy of nuclei is high enough for close interactions to occur. Even then, such interactions must be sufficiently frequent that we can "pay back" the cost of producing the high
temperature plasma in the first place. Thus, the plasma must be
kept hot for a sufficiently long time τ, <u>the containment time</u>, and
so must be a very stable object. Apart from the containment time
against various types of instability, the plasma is prone to other,
inevitable, sources of energy loss. Thus, high energy neutral
particles, which may be produced either in primary fusion reactions,
or subsequently by charge exchange collisions, escape immediately
from the plasma. Radiation loss (Bremsstrahlung) may also lead to
serious energy loss. In both cases, effects can be minimised by
absorption and reflection outside the plasma region.

We see therefore that the temperature T and the containment time
τ play a vital role. The third important plasma property is the
particle density n, for binary collisions will increase as n^2.
The appropriate criterion, known as the <u>Lawson criterion</u>,[3] is
truly "astronomical". For a deuterium gas, the temperature must
exceed 10^8K and n, τ combine in the inequality

$$n\tau > 10^{22} \text{ m}^{-3} \text{ s}.$$

2

For a gas composed of a 50:50 mixture of deuterium and tritium, the temperature must exceed only 10^7K, while the density, containment time criterion is

$$n\tau > 10^{20} \text{ m}^{-3} \text{ s.}$$

During the past few years, attempts to achieve this value of $n\tau$ have concentrated on experiments devised on the basis of magnetic field confinement. As we shall see later, a high temperature plasma is a good electrical conductor, so that a large current density $\underline{j}$ and magnetic field $\underline{B}$ can be generated. Magnetic confinement depends on the body force $\underline{j} \times \underline{B}$ being directed away from the walls of the containing vessel. An immediate consequence of the low electrical resistivity of the plasma is the conservation of magnetic flux in any plasma motion, leading to the concept of a magnetic pressure acting on the plasma. In the current series of Tokamak toroidal plasma experiments, initiated in the U.S.S.R., and now widely accepted as one positive road towards a nuclear fusion reactor, a density of 10^{19} m^{-3} and containment time $\tau \simeq 10^{-1}$ s, gives a value for $n\tau$ of 10^{18} m^{-3} s. Temperatures in excess of 10^7K for the electron component of the plasma, and somewhat less for ions, taken together with the quoted $n\tau$ value, suggest a very encouraging future for this line of research. Indeed, plans are being laid at present in many countries to invest in large research programmes specifically to test out fully the potential of this approach.

More recently, a new possibility has been under investigation, initially by computer simulation, but now also experimentally, which suggests that a small pellet (of radius 1mm) of solid deuterium (or d-t mixture) can be compressed, by means of high power laser beams arranged symmetrically, and fired simultaneously, to provide spherical implosion to more than 10^4 times solid density, and at a final temperature around 10^8K. Preliminary calculations indicate that the total laser energy required, so that the thermonuclear energy output should just balance the laser energy input, is 1kJ in a 1 ns pulse. We may note in passing that the final plasma pressure exceeds 10^{12} atmospheres, and that electrons are Fermi degenerate, so that a new term, superdense plasma, has been introduced to describe this state, perhaps more akin to the nature of a white dwarf star in _some_ of its properties, and certainly well worth investigating in its own right.

In such a compression, then, we should expect $n \simeq 5 \times 10^{32}$ m^{-3},

and τ can be taken as the inertial time, or the time for the
pellet to blow apart. To obtain an estimate for τ, we note that
the sound speed is of order 10^6 ms^{-1} and the final pellet radius of
order 10^{-4} m. Hence $\tau \simeq 10^{-10}$ s and $n\tau \simeq 5 \times 10^{22}$. The Lawson
criterion should therefore in principle be relatively easy to
satisfy.

While the application to thermonuclear research is of paramount
importance, other technological applications are also being explored,
which will receive only very brief mention here. One is the
direct conversion of the kinetic energy of a fast flowing plasma
into electrical energy. A second is the possibility of propulsion
by an ion jet. This latter application may be of great importance
one day for long-range space flights, for it is the exhaust velocity
of the jet which ultimately determines the terminal velocity of the
space ship. In chemically-driven rockets this limit has virtually
been reached, but an ion jet, being accelerated by electromagnetic
forces, can in principle attain a speed comparable with the speed
of light.

1-2 BASIC PLASMA PROPERTIES

Our next task is to derive from basic principles certain properties
of a fully-ionised quiescent plasma, which appear frequently in the
more rigorous and involved treatment of plasma processes which are
discussed subsequently. The simple derivations given here do not
necessarily lack rigour; but the physical and geometrical complexities are kept to a minimum so that the essential features can
easily be distinguished.

1-2-1 Screening length λ

A plasma tends to be electrically neutral, for if over a large
volume the electron density were to deviate appreciably from the ion
density, a large electrostatic potential would develop, and the
resultant electrostatic forces would tend to restore charge neutrality. Nevertheless, in a high temperature plasma, thermal motion
of ions and electrons will lead to density fluctuations. Assuming
that ions and electrons are in thermal equilibrium at a temperature
T, so that, on average, the kinetic energy of a particle is $\frac{3}{2}\kappa T$,
electrons will move about randomly much more rapidly than ions.
This thermal motion will allow a certain departure from complete,
microscopic, charge neutrality.

We should expect an enhancement of the electron density near a particular ion, so that, if we measure the electrostatic field of the ion, we should expect it to fall off as r^{-2} if we are sufficiently close (no screening electrons present), but with increasing distance r the ion field will be screened off by the surrounding excess of electrons. We shall see that there is a characteristic shielding distance λ, which increases with temperature but decreases with density. We shall demonstrate this plasma property firstly by using a simple slab model of a plasma.

Consider a model situation in which plasma fills space uniformly except in a slab extending from $-d$ to $+d$, where only ions are present, with uniform density n, as shown in Fig. 1.1.

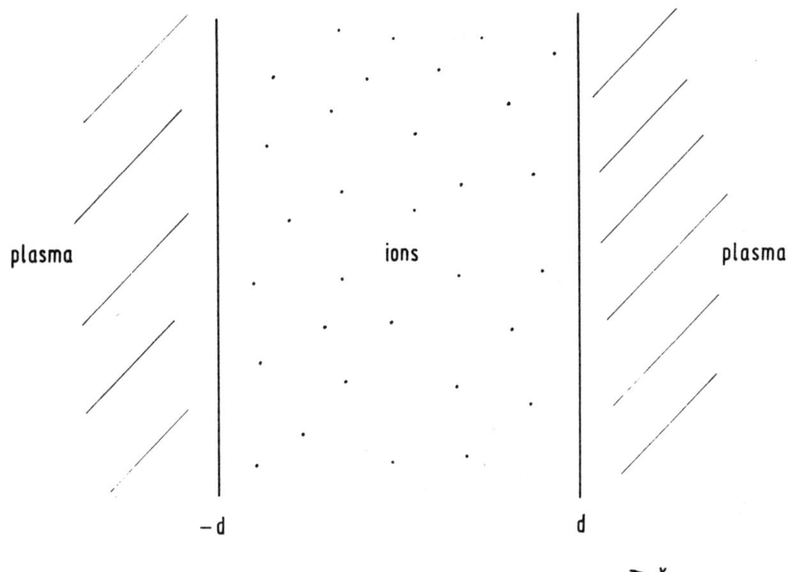

Fig. 1.1 Model for charge screening

There will be an electrostatic potential ϕ in the ion region, $\phi = 0$ in the rest of the plasma. ϕ satisfies Poisson's equation

$$\varepsilon_o \frac{d^2\phi}{dx^2} = -ne$$

with solution $\phi(x) = ne(d^2 - x^2)/2\varepsilon_o$.

We now suppose that this ion region can be physically maintained by thermal fluctuations of the electron density. In particular, we shall suppose that an electron at $x = 0$ initially, with average kinetic energy $\tfrac{1}{2}\kappa T$ associated with its motion parallel to the x-axis, will just manage to overcome the potential barrier due to the ions. We thus equate $\tfrac{1}{2}\kappa T$ to the maximum electron potential energy $e\phi(0)$:

$$\tfrac{1}{2}\kappa T \;=\; \frac{ne^2\,d^2}{2\varepsilon_o}$$

$$\text{i.e.}\qquad d^2 \;=\; \frac{\varepsilon_o \kappa T}{ne^2}\;.$$

The length d is identified with the characteristic screening
length λ, so that

$$\lambda \;=\; \left(\frac{\varepsilon_o \kappa T}{ne^2}\right)^{\tfrac{1}{2}}\;. \tag{1.1}$$

Statistical mechanics provides an elegant way of obtaining a
similar result.   In a canonical ensemble of interacting particles,
the distribution of particles in phase space is given by $f \simeq$
$\exp(-H/\kappa T)$, where H is the Hamiltonian function.   For weakly
interacting particles, $H = \tfrac{1}{2}mv^2$ and we obtain the usual Maxwell-
Boltzmann velocity distribution.   For particles interacting
through a two-body potential V we get $H = \tfrac{1}{2}mv^2 + V(\underline{r})$, and the
Boltzmann distribution is $f \simeq \exp(-V/\kappa T)\exp(-\tfrac{1}{2}mv^2/\kappa T)$, where V
may vary spatially.   This is just the case for electrons surround-
ing an ion.   They are attracted to it by some effective potential,
Coulombic only close to the ion.   Now since f tells us the number
of electrons with velocity $\underline{v}$ at a distance $\underline{r}$ from the ion,
$\exp(-V(\underline{r})/\kappa T)$ will tell us how many electrons of all velocities
are at distance $\underline{r}$, and $V = -e\phi$.   Thus

$$n_e(\underline{r}) \;=\; n \,\exp\,(e\phi(\underline{r})/\kappa T),$$

where n_e is the electron density, now normalised at large distances,
where $\phi \to 0$, to the background ion density n.

　　We shall assume that $e\phi \ll \kappa T$ over most of space, which is
usually the case.

　　Thus,　　　　　$n_e(\underline{r}) \;\simeq\; n\,(1 + e\phi/\kappa T).$

$\phi(\underline{r})$ satisfies Poisson's equation

$$\nabla^2\phi \;=\; -\,(n_i - n_e)\,\frac{e}{\varepsilon_o}$$

$$\;=\; \frac{ne^2\phi}{\varepsilon_o \kappa T}$$

$$\;=\; \frac{\phi}{\lambda^2}\;. \tag{1.2}$$

The same characteristic length appears again.   Assuming spherical
symmetry about the ion,

$$\nabla^2\phi \;\to\; \frac{1}{r^2}\,\frac{d}{dr}\left(r^2\,\frac{d\phi}{dr}\right).$$

Boundary conditions are

$$\phi \;\to\; \frac{e}{4\pi\varepsilon_o r}\;, \qquad r \to 0 \qquad \text{(Coulomb)}$$

$$\phi \;\to\; 0, \qquad\qquad r \to \infty \qquad \text{(complete shielding)}$$

leading to a solution

$$\phi \;=\; \frac{e}{4\pi\varepsilon_o r}\;\exp\left(-\frac{r}{\lambda}\right). \tag{1.3}$$

Thus ϕ falls off exponentially over a characteristic distance λ.
This theory was in fact first applied to the motion of charges in
electrolytes by Debye, and λ is often termed the <u>Debye Length</u>.

Table 1.1 gives λ for various values of density and temperature.

$n(m^{-3})$ / $T(eV)$	10^{16}	10^{18}
10	$2\cdot3 \times 10^{-4}$	$2\cdot3 \times 10^{-5}$
100	$7\cdot4 \times 10^{-4}$	$7\cdot4 \times 10^{-5}$

Table 1.1 Screening length λ (metres).

It is interesting to obtain λ for two widely different situations.

(i) Tokamak plasma

$$n \;=\; 10^{20}\ m^{-3}$$
$$T \;=\; 1\ keV$$
$$\lambda \;\simeq\; 2 \times 10^{-5}\ m$$

(ii) hot interstellar gas

$$n \;=\; 10^{6}\ m^{-3}$$
$$T \;=\; 1\ eV$$
$$\lambda \;\simeq\; 7\cdot4\ m$$

An important non-dimensional parameter is $N = n\lambda^3$, a measure of the
number of particles in a sphere of radius λ. We would normally
require N to be a large number in order to have a plasma at all.
In the two cases above, (i) $N \simeq 10^6$ and (ii) $N \simeq 4 \times 10^8$.

1-2-2 Plasma frequency ω_p

The next elementary property we shall study is that of plasma
oscillations. Our first conclusion, that a plasma is electrically
neutral over distances greater than λ, establishes also the existence of restoring forces, should charge neutrality be disturbed.
We consider the simplest situation in which thermal effects are
completely ignored. Initially, the plasma is spatially uniform,
with equal densities n_o of ions and electrons. This steady state
is locally disturbed by an increase δn in the electron density.
Because ions are much more massive than electrons, we shall assume
that they simply supply a positive neutralising background, but
play no active part in the plasma dynamics. We shall this time

describe the electrons as though they formed a fluid.

(i) Equation of continuity

The net increase in density in a volume element equals the net flux across the surface of the element. Hence

$$\frac{\partial n}{\partial t} + \nabla \cdot (n\underline{v}) = 0, \tag{1.4}$$

where $n = n_o + \delta n$

$\quad \underline{v} =$ average electron velocity

$\quad\quad \equiv$ electron fluid velocity.

(ii) Poisson's equation

$$\varepsilon_o \nabla \cdot \underline{E} = -e (n - n_o) = -e\delta n, \tag{1.5}$$

since the ion density n_o is constant.

(iii) Equation of motion

The rate of change of momentum of a volume element is equal to the total force acting on the element.

$$mn \frac{d\underline{v}}{dt} = -ne\underline{E}. \tag{1.6}$$

Initially, the electrons are at rest, so that $\underline{v}$ and $\underline{E}$ are both initially zero. If $\delta n / n_o \ll 1$, we may take $\underline{v}$ and $\underline{E} \simeq 0(\delta n)$, and, to first order, equations (1.4) - (1.6) become

$$\frac{\partial}{\partial t} (\delta n) + n_o \nabla \cdot \underline{v} = 0$$

$$\nabla \cdot \underline{E} = -\frac{e}{\varepsilon_o} \delta n$$

$$m \frac{\partial \underline{v}}{\partial t} = -e\underline{E}.$$

Elimination of $\underline{v}$ and $\underline{E}$ yields an equation for δn:

$$\left[\frac{\partial^2}{\partial t^2} + \frac{n_o e^2}{m \, \varepsilon_o} \right] \delta n = 0. \tag{1.7}$$

Thus, δn oscillates about its mean value n_o with frequency

$$\omega_p = \left(\frac{n_o e^2}{m \, \varepsilon_o} \right)^{\frac{1}{2}}.$$

Note that in the derivation given above we have glossed over an important point. The velocity $\underline{v}$ is in effect the velocity vector field $\underline{v}(\underline{r}, t)$ of the electron fluid. Hence, $\frac{d\underline{v}}{dt} \equiv \frac{\partial \underline{v}}{\partial t} + (\underline{v} \cdot \nabla)\underline{v}$. However, since $\underline{v}$ is a first-order quantity, $(\underline{v} \cdot \nabla)\underline{v}$ is second-order and may be neglected.

1-2-3 Dielectric coefficient ε

The passage of an electromagnetic wave through a plasma induces an electron current, the effect of which can be interpreted in terms of dielectric properties.

For a wave of frequency ω, let the electric field vector be

$$\underline{E}(t) = \underline{E}_o \exp(i\omega t).$$

The velocity of electrons is given by

$$\frac{d\underline{v}}{dt} = -\frac{e}{m}\,\underline{E},$$

which gives $\underline{v} = -\dfrac{e}{i\omega m}\,\underline{E}$, and an induced current

$$\underline{j} = -ne\underline{v} = \frac{ne^2}{i\omega m}\,\underline{E}. \tag{1.8}$$

The effect of this induced current can be seen by using Maxwell's equation

$$\nabla \times \underline{H} = \underline{j} + \frac{\partial \underline{D}}{\partial t} \tag{1.9}$$

$$= \left(\frac{ne^2}{i\omega m} + i\omega\varepsilon_o\right)\underline{E}$$

$$= i\omega\varepsilon_o\left(1 - \frac{\omega_p^2}{\omega^2}\right)\underline{E}.$$

Hence the plasma behaves towards an electromagnetic wave of frequency ω as a dielectric with coefficient

$$\varepsilon = \varepsilon_o\left(1 - \frac{\omega_p^2}{\omega^2}\right);$$

and notice that it is frequency-dependent, with no propagation for $\omega < \omega_p$.

The effect of a constant magnetic field may be obtained in a similar way. Let us suppose the field to be directed parallel to the x-axis of a set of co-ordinate axes. Thus, $\underline{B} = B\,\hat{\underline{x}}$. Then, the equation of motion for electrons is

$$\frac{d\underline{v}}{dt} = -\frac{e}{m}(\underline{E} + \underline{v} \times \underline{B}). \tag{1.10}$$

In component form, we have

$$\frac{dv_x}{dt} = -\frac{e}{m}E_x$$

$$\frac{dv_y}{dt} = -\frac{e}{m}E_y - \Omega v_z$$

$$\frac{dv_z}{dt} = -\frac{e}{m}E_z + \Omega v_y, \tag{1.11}$$

where $\Omega = eB/m =$ electron gyrofrequency.

The solution for $\underline{v}$, varying as exp $(i\omega t)$ in response to $\underline{E}$ as before, is

$$v_x = \frac{ie}{m\omega} E_x$$

$$v_y = \frac{ie}{m} \frac{\omega E_y + i\Omega E_z}{\omega^2 - \Omega^2}$$

$$v_z = \frac{ie}{m} \frac{\omega E_z - i\Omega E_y}{\omega^2 - \Omega^2} . \tag{1.12}$$

Thus, a current $\underline{j}$ is set up, which in the case of a magnetised plasma is not parallel to $\underline{E}$:

$$j_x = -i\omega\varepsilon_o \frac{\omega_p^2}{\omega^2} E_x$$

$$j_y = -i\omega\varepsilon_o \omega_p^2 \left[E_y \frac{1}{\omega^2 - \Omega^2} + iE_z \frac{\Omega}{\omega(\omega^2 - \Omega^2)} \right]$$

$$j_z = -i\omega\varepsilon_o \omega_p^2 \left[E_z \frac{1}{\omega^2 - \Omega^2} - iE_y \frac{\Omega}{\omega(\omega^2 - \Omega^2)} \right] . \tag{1.13}$$

The dielectric coefficient is now a tensor $\underline{\underline{\varepsilon}}$ defined by

$$\underline{\underline{\varepsilon}} . \underline{E} = \underline{D} = \varepsilon_o \underline{E} + \frac{\underline{j}}{i\omega} . \tag{1.14}$$

Thus, $\underline{\underline{\varepsilon}} = \varepsilon_o \begin{bmatrix} 1 - \dfrac{\omega_p^2}{\omega^2} & 0 & 0 \\[3mm] 0 & 1 - \dfrac{\omega_p^2}{\omega^2 - \Omega^2} & \dfrac{-i\omega_p^2 \Omega}{\omega(\omega^2 - \Omega^2)} \\[3mm] 0 & \dfrac{i\omega_p^2 \Omega}{\omega(\omega^2 - \Omega^2)} & 1 - \dfrac{\omega_p^2}{\omega^2 - \Omega^2} \end{bmatrix} . \tag{1.15}$$

This of course reduces to the earlier result for an unmagnetised plasma, if $B = 0$, that is, $\Omega = 0$. Physically, this result corresponds to the ordinary plasma longitudinal mode obtained in the absence of a magnetic field, together with a mixture of right and left circularly polarised transverse waves.

It is of interest to estimate the strength of the magnetic field required before the dielectric coefficient departs significantly from the unmagnetised plasma coefficient. A simple criterion is $\Omega^2 > \omega_p^2$, which can be written

$$B^2 > \mu_o nmc^2 .$$

For $n = 10^{20}$ m^{-3}, B > 3 tesla.

These are typically Tokamak parameters.

1-2-4 Electrical resistivity η

Since a plasma contains essentially free electrons and ions, it
might be expected to be a good electrical conductor. We shall
assume that a form of Ohm's law is obeyed, so that $\eta = E/j$.
Here E is a constant electric field applied across the plasma,
and j the resulting steady current density. This assumption will
be reasonably valid provided E is a weak field. We view the
process of electrical conduction at the particle level. When E
is applied, electrons flow in one direction, ions in the opposite
direction, though once more we shall neglect ion motion. As a
result of charge interactions, momentum is exchanged between ions
and electrons until a steady state is reached, when the rate of
change of momentum of electrons is exactly balanced by the force
due to the applied electric field.

Let q be the rate at which a single electron transfers its
momentum in collisions with ions. Then q = eE. The electron
current is j = nev, where v is the average electron drift velocity,
n the electron density. Then

$$\eta = \frac{E}{j} = \frac{q}{ne^2v} . \tag{1.16}$$

So far, the problem has just been restated, so that now, instead
of finding j for an applied E, we are to find v for a given q.
Next, we turn the problem round and find q for a given v. We
shall show that q is proportional to n, and evaluate the coefficient
of proportionality, which will yield the electrical resistivity η.

Suppose an electron is deflected through an angle θ by an ion.
Initially its momentum is parallel to E and equal to mv, but after
the collision the component of momentum parallel to E is mv cos θ.
Hence, the momentum transfer parallel to E is

$$mv (1 - \cos \theta).$$

The Rutherford differential cross section is given by

$$\frac{d\sigma}{d\Omega} = \left(\frac{e^2}{8\pi\varepsilon_o mv^2} \right)^2 \mathrm{cosec}^4 \tfrac{1}{2}\theta.$$

The rate of transfer of momentum q can now be stated in the form

$$q \;=\; nv \int_{\theta_{min}}^{\theta_{max}} mv \, (1 - \cos\theta) \, \frac{d\sigma}{d\Omega} \, 2\pi \sin\theta \; d\theta, \qquad (1.17)$$

noting that nv is the number of scattering centres per unit area per second encountered by an electron of velocity v, and assuming all scattering angles are equally probable in the range (θ_{min}, θ_{max}). We first note that θ_{max} may be taken equal to π but that q is divergent at $\theta = 0$, for

$$\frac{1}{8} \int (1 - \cos\theta) \, \mathrm{cosec}^2 \tfrac{1}{2}\theta \, \sin\theta \; d\theta$$

$$= \int \frac{\cos \tfrac{1}{2}\theta}{\sin \tfrac{1}{2}\theta} \, d(\tfrac{1}{2}\theta) \;=\; \ell n \, \sin \tfrac{1}{2}\theta .$$

The problem of divergence can be resolved, if we recall that the path of an electron in a Coulomb force field is a hyperbola. We can show that

$$\mathrm{cosec}^2 \tfrac{1}{2}\theta \;=\; 1 + \left(\frac{4\pi\varepsilon_o \, mv^2 b}{e^2} \right)^2$$

for an electron of velocity v, scattered through an angle θ with impact parameter b. We then make the assumption that, if $b > \lambda$, the electron is everywhere outside the <u>screened</u> Coulomb field of the ion and is unaffected by it. We further assume that

$$\lambda \;\gg\; \frac{e^2}{4\pi\varepsilon_o \, mv^2} \;,$$ for v around thermal speed. Then θ_{min} is defined by taking $b = \lambda$, that is

$$\mathrm{cosec} \, \tfrac{1}{2}\theta_{min} \;\simeq\; \frac{\lambda}{e^2/4\pi\varepsilon_o \, mv^2} \;.$$

Substituting into the expression for q, we obtain

$$q \;=\; \frac{ne^4}{4\pi\varepsilon_o^2 mv^2} \; \ell n \; \mathrm{cosec} \, \tfrac{1}{2}\theta_{min} . \qquad (1.18)$$

Strictly, we should average the expression for q over a Boltzmann distribution exp $(-\tfrac{1}{2}mv^2/\kappa T)$, but it will be sufficient to take $mv^2 \simeq 3\kappa T$. Also, if we define

$$\Lambda \;=\; \frac{\lambda}{e^2/4\pi\varepsilon_o \, mv^2} \;\simeq\; (4\pi\varepsilon_o)^{\frac{3}{2}} \; \frac{3}{2e^3} \; \left(\frac{\kappa^3 T^3}{\pi n} \right)^{\frac{1}{2}} ,$$

we arrive at an expression for the resistivity

$$\eta \simeq \frac{m^{\frac{1}{2}} e^2 \ln \Lambda}{4\pi (3\kappa T)^{\frac{3}{2}} \varepsilon_o^2} \,, \qquad\qquad (1.19)$$

(i)   Note that η is independent of n.

(ii)   Since $\ln \Lambda$ is a slowly varying function of T, η exhibits a temperature dependence $T^{-\frac{3}{2}}$.

(iii)   At $T = 10^7 K$, $n = 10^{21}$ electrons/m^3 a plasma has a resistivity which approximates to that of copper (at room temperature).

Chapter 2

ORBIT THEORY[4]

2-1 INTRODUCTION

The study of the motion of a single charged particle in given electric and magnetic fields is an important branch of plasma physics, since many plasma properties can be understood from a single particle description.

First, let us consider the simple case of the motion of a charged particle in a uniform magnetic field $\underline{B}$. The path of the particle is a helix with axis parallel to the magnetic field. The angular frequency of rotation of the particle about the field line is the gyrofrequency $\Omega = \dfrac{eB}{m}$, and if its component of velocity perpendicular to the magnetic field is $v_\perp$ then its radius of gyration $r_o = v_\perp / \Omega$. Its component of velocity $v_\parallel$ parallel to the magnetic field is unaffected by the field and, like $v_\perp$, is strictly a constant of the motion.

Suppose now that the magnetic field varies slowly in space and time. We need some criterion for slow variation. We take it to be

$$\frac{1}{\Omega B}\frac{\partial B}{\partial t} \ll 1; \qquad \frac{r_o}{B}\frac{\partial B}{\partial x} \ll 1$$

which means simply that changes in B are very small

(i) in one period of rotation $\dfrac{2\pi}{\Omega}$;

(ii) over a distance equal to the radius of gyration r_o.

We shall find that in slowly varying fields the motion can be considerably modified. Orbits are almost helical, but because the motion is no longer exactly periodic, the motion along $\underline{B}$ is coupled to motion perpendicular to the field. One important feature is the existence of adiabatic invariants of the motion. In general, an adiabatic invariant is a quantity whose change approaches zero as some physical parameter approaches some limiting value. The one of most significance to the study of plasmas is the magnetic moment μ associated with gyration about field lines. A consequence of this approximate constant of motion is the slow drift of a

particle across the field. We can represent the motion of the
particle in terms of a slowly varying 'guiding centre' trajectory
accompanied by a rapid gyration about the guiding centre motion.
The latter, in the case of a uniform magnetic field, coincides
with a field line, but if the field varies the guiding centre
drifts away from the field line. This will be the topic of
§2-3. In the final section some simple applications of the
results are presented.

2-2 ADIABATIC INVARIANTS

Consider first the case of a uniform magnetic field. The
particle moves in a circle of radius r_o and angular velocity Ω in
projection on a plane normal to the magnetic field. It therefore
constitutes a circular current I enclosing an area A, with

$$I = \frac{e\Omega}{2\pi} \text{ , for a particle of charge e}$$

$$A = \pi r_o^2 .$$

Hence, the magnetic moment μ associated with the gyration of the
particle about a field line is

$$\mu = AI$$

$$= \pi r_o^2 \frac{e\Omega}{2\pi}$$

$$= \frac{\tfrac{1}{2}mv_\perp^2}{B} .$$

Expressing the kinetic energy W of the particle as a sum

$$W = \tfrac{1}{2}mv_\perp^2 + \tfrac{1}{2}mv_\parallel^2 = W_\perp + W_\parallel$$

we have

$$\mu = \frac{W_\perp}{B} .$$

We demonstrate the invariance of μ in two special cases:

(i) Spatial variation of $\underline{B}$

Since the force on the particle is $e\underline{v} \times \underline{B}$, it is always perpendicular to the direction of motion. Hence, the angular momentum
will remain approximately constant. Thus

$$mv_\perp r_o = \text{constant}$$

$$\text{i.e. } \frac{mv_\perp^2}{\Omega} = \text{constant}$$

and LHS is proportional to μ.

We therefore expect μ to remain constant. The same result can be obtained by another argument in the special case of axial symmetry. Taking the z-axis as the axis of symmetry, and introducing cylindrical coordinates (r, θ, z), the condition that the magnetic field should be divergence-free, $\nabla . \underline{B} = 0$, is

$$\frac{1}{r} \frac{\partial}{\partial r} (rB_r) + \frac{\partial B_z}{\partial z} = 0.$$

Restricting the particle motion to lie close to the axis, and taking $B_z \simeq |\underline{B}| = B(z)$ there, we get an approximate form for B_r

$$B_r \simeq -\tfrac{1}{2}r \frac{dB}{dz} .$$

The equation of motion is

$$\frac{d\underline{v}}{dt} = \frac{e}{m} \underline{v} \times \underline{B}, \tag{2.1}$$

and we note that, if we set $\underline{\Omega} = - \frac{e\underline{B}}{m}$, (2.1) becomes

$$\frac{d\underline{v}}{dt} = \underline{\Omega} \times \underline{v} ,$$

showing that the angular velocity of the particle motion is directed anti-parallel to $\underline{B}$ for positive charge.

Taking the z-component of (2.1) gives

$$\frac{dv_z}{dt} = - \frac{e}{m} v_\theta B_r = \frac{e}{2m} rv_\theta \frac{dB}{dz} .$$

We consider only the case in which the orbit encircles the axis and we assume that $r \simeq r_o$ and $v_\theta \simeq -v_\perp$. We then obtain

$$\frac{dv_z}{dt} = - \frac{er_o v_\perp}{2m} \frac{dB}{dz} = - \frac{\mu}{m} \frac{dB}{dz} . \tag{2.2}$$

In the same approximation, $\frac{d}{dt} \simeq v_z \frac{d}{dz}$, so that $\frac{dW_z}{dt} = -\mu \frac{dB}{dt}$, following the path of the particle and setting $W_z = \tfrac{1}{2}mv_z^2$. Since no work is done by the magnetic field on the particle, $W_z + W_\perp =$ constant. Hence

$$\frac{dW_z}{dt} = - \frac{dW_\perp}{dt} = - \frac{d}{dt} (\mu B),$$

and thus $\frac{d}{dt} (\mu B) = \mu \frac{dB}{dt}$ showing that μ is conserved in this approximation.

(ii) Time variation of $\underline{B}$

Since $\underline{B}$ is now varying with time, an e.m.f. $\mathcal{E}$ will be induced around the particle orbit.

$$\mathcal{E} = \oint \underline{E} . \underline{ds} = -\int \frac{\partial \underline{B}}{\partial t} . \underline{dA} \simeq - \frac{dB}{dt} \pi r_o^2.$$

This e.m.f. does not affect W_z, the kinetic energy associated
with motion parallel to the magnetic field.   However

$$\frac{dW_\perp}{dt} \; = \; \mathscr{E} I$$

$$\text{i.e. } \frac{d}{dt}(\mu B) \; = \; \frac{e\Omega}{2\pi} \; \pi r_o^2 \; \frac{dB}{dt}$$

noting that the current and e.m.f. are in the same direction if B
is increasing with time.   Hence $\frac{d}{dt}(\mu B) = \mu \frac{dB}{dt}$, again implying
conservation of μ.

We now sketch briefly a more general proof of adiabatic
invariance of μ, based on the theory of action integrals in
analytical dynamics.   Adiabatic invariants in general arise, in
systems with several degrees of freedom, if the frequencies
associated with oscillations corresponding to these degrees of
freedom differ by orders of magnitude.   For a particle moving in
a strong magnetic field which varies slowly in space and time
according to the criteria set down in §2-1, the three degrees of
freedom can be taken approximately as (a) gyration about a field
line, (b) motion along the field line, and (c) drift across the
field.   It proves to be possible in certain circumstances to
decouple the motion in each degree of freedom, in which case the
adiabatic invariants are just the action integrals corresponding
to each individual degree of freedom.   The magnetic moment is
then proportional to the action integral associated with gyration
about a field line.

The position vector $\underline{r}(t)$ of a particle is decomposed into two
parts

$$\underline{r}(t) \; = \; \underline{R}(t) + \underline{\xi}(t), \tag{2.3}$$

where $\underline{R}$ is the position of the guiding centre and $\underline{\xi}$ is the rapidly
varying part corresponding to gyrations about the field.   Taking
the time derivative of (2.3), we get

$$\underline{v} \; = \; \underline{u} + \underline{w},$$

$\underline{u}$ corresponding to the guiding centre velocity and $\underline{w}$ to the
velocity of gyration.

We now assume that $\underline{v}_\parallel = \underline{u}_\parallel$ and $\underline{v}_\perp \simeq \underline{w}$ which implies that the
perpendicular drift velocity $\underline{u}_\perp$, which we are particularly
interested in calculating, is much less than $\underline{w}$.

In Hamiltonian formalism, the generalised momentum $\underline{p}$ for a
charged particle moving in a magnetic field is given by

$$\underline{p} = m\underline{v} + e\underline{A},$$

so that for motion perpendicular to $\underline{B}$ we have

$$\underline{p}_\perp = m\underline{w} + e\underline{A}_\perp.$$

Our assumptions above imply that gyrations are closed in lowest order, so that the action integral

$$\oint \underline{p}_\perp \cdot d\underline{l} = \text{constant}, \qquad (2.4)$$

which can be written, noting that $d\underline{l}$ is perpendicular to $\underline{B}$, as

$$\oint m\underline{w}.d\underline{l} + \oint e\underline{A}.d\underline{l} = \text{constant}. \qquad (2.5)$$

In (2.5), $m\underline{w} = e\underline{r} \times \underline{B}$ so that the first term

$$\oint m\underline{w}.d\underline{l} = e \oint \underline{r} \times \underline{B}.d\underline{l} \simeq -2\pi r^2 \, eB,$$

where B has been assumed uniform over the particle gyration orbit, in accordance with our criteria. The second term in (2.5) can be evaluated thus:

$$e\oint \underline{A}.d\underline{l} = e\int \nabla \times \underline{A}.d\underline{S} = e\int \underline{B}.d\underline{S} = \pi r^2 \, eB.$$

Since both terms have the same form (and clearly do not cancel each other out), the action integral invariant (2.4) reduces to a statement of flux conservation, namely, the total magnetic flux enclosed by the orbit is a constant of the motion. Now, the magnetic moment μ is just $\frac{e}{m}$ times this magnetic flux. We have thus recovered quite generally the result obtained previously in special cases.

Having obtained the constant of the motion for the degree of freedom with highest oscillation frequency, we can decouple that motion from the other two degrees of freedom, since we can express $u_{\parallel}$ as a function of position s measured along the magnetic field line in question. For,

$$u_{\parallel} = \left[\frac{2}{m} \left(W - \mu B(s) \right) \right]^{\frac{1}{2}}.$$

If $B(s)$ varies such that $W = \mu B$ somewhere along the path, then $u_{\parallel} = 0$, and the particle is reflected. This is the principle of the magnetic mirror. Consider now a particle oscillating between two such reflection points, such that the time between successive reflections is much longer than the average gyromagnetic period $\frac{2\pi}{\Omega}$. We can then define a second invariant in terms of the action integral for motion along the field line,

$$\oint p_{\parallel} \, ds = \text{constant}.$$

Since no magnetic flux is enclosed in the complete oscillation,
which is from one reflection point to the other and back, the
action integral reduces to the simpler form

$$J = \oint u_{\parallel} \, ds$$

$$= \oint \left[\frac{2}{m} \left(W - \mu B(s) \right) \right]^{\frac{1}{2}} \, ds.$$

We shall show in the next section that the presence of gradients
in the magnetic field causes the particle to drift to neighbouring
field lines.   It sometimes happens that this drift motion is also
periodic.   (This can occur for example if there is an axis of
symmetry in the system and the drift motion encircles the axis.)
We then expect to obtain a third invariant.   By analogy with the
discussion which led to magnetic flux conservation in the case of
gyration about field lines, we now find a flux invariant of the
form

$$\Phi = \int_{S} \underline{B} . d\underline{S} = \text{constant},$$

where S is encircled by the drift orbit.

2-3 GUIDING CENTRE DRIFT MOTION

In the previous section, equation (2.3), the position vector $\underline{R}$ of
the guiding centre was introduced.   We now derive an equation of
motion for $\underline{R}$, for which the rapid gyration about the magnetic
field, described by $\underline{\xi}$, has been averaged out.   We have of course
already obtained an insight into this motion, since it is
associated with the longitudinal invariant J and the flux invariant
Φ.   To make further progress in studying the drift motion we start
with the exact equation of motion (2.1), in the case of zero
electric field;   the effect of electric field is considered later.

In this equation, $\underline{B}$ is a function of $\underline{r}$ and t.   We now express
B as a function of R and $\underline{\xi}$ in the neighbourhood of the position of
the guiding centre, retaining only the first power in $\underline{\xi}$ in a
Taylor series about $\underline{R}$.   Thus,

$$\underline{B}(\underline{r}) = \underline{B}(\underline{R}) + (\underline{\xi} . \nabla) \, \underline{B}(\underline{R}).$$

We also express $\underline{v}$ as $(\underline{u} + \underline{w})$ and obtain

$$\frac{d\underline{u}}{dt} + \frac{d\underline{w}}{dt} = \frac{e}{m} (\underline{u} + \underline{w}) \times (\underline{B} + (\underline{\xi} . \nabla) \, \underline{B}), \qquad (2.6)$$

where $\underline{B}$ stands for $\underline{B}(\underline{R})$.

Equation (2.6) will now be written with all terms on RHS oscillating with the gyrofrequency and averaging to zero to first order.

$$\frac{d\underline{u}}{dt} - \frac{e}{m} \underline{u} \times \underline{B} - \frac{e}{m} \underline{w} \times (\underline{\xi}.\nabla)\underline{B}$$

$$= - \frac{d\underline{w}}{dt} + \frac{e}{m} \underline{u} \times (\underline{\xi}.\nabla)\underline{B} + \frac{e}{m} \underline{w} \times \underline{B}.$$

Note that the third term on LHS involves a product of $\underline{w}$ and $\underline{\xi}$ and has a non-zero average. Averaging over a gyroperiod, we get

$$\frac{d\underline{u}}{dt} - \frac{e}{m} \underline{u} \times \underline{B} - \frac{e}{m} \overline{\underline{w} \times (\underline{\xi}.\nabla)\underline{B}} = 0. \qquad (2.8)$$

In equation (2.8), we have to obtain an expression for $\overline{\underline{w} \times (\underline{\xi}.\nabla)\underline{B}}$, averaging over a gyroperiod. If we set $\Omega\underline{\xi} = \underline{w}$, note that to lowest order $\underline{w}$ has a time dependence $\cos \Omega t$, and write θ for Ωt, we have

$$\overline{\underline{w} \times (\underline{\xi}.\nabla)\underline{B}} = -\nabla B \frac{1}{2\pi\Omega} \oint d\theta \, w^2 \cos^2 \theta$$

$$= - \frac{\mu}{e} \nabla B.$$

The equation for the guiding centre is thus

$$m \frac{d\underline{u}}{dt} = e \, \underline{u} \times \underline{B} - \mu\nabla B, \qquad (2.9)$$

with the reminder that $\underline{B}$ is evaluated at the guiding centre, not at the location of the particle. Note that equation (2.9) is a generalisation of (2.2), which was obtained as a special case where the magnetic field possesses an axis of symmetry and the guiding centre is moving along that axis. It is useful to view the term $- \mu\nabla B$ as a force acting on the particle, resulting from a potential μB. There is obvious similarity between this quantity and the potential energy $- \underline{\mu}.\underline{B}$ of a permanent magnetic dipole of moment $\underline{\mu} = - \frac{\mu\underline{B}}{B}$ in a magnetic field $\underline{B}$, noting that the magnetic moment is antiparallel to $\underline{B}$, as we saw in the previous section. The analogy should not be taken too literally.

Equation (2.9) describes the complete guiding centre motion in a magnetic field. We are often interested in the magnetostatic case, for which it is instructive to obtain separate equations for the components of $\underline{u}$ parallel and perpendicular to $\underline{B}$. This shows more clearly the various physical effects which contribute to the drift motion.

We make use of the fact that to lowest order $\underline{u} = u_{\parallel} \underline{b}$, where

$\underline{b}$ is a unit vector parallel to the magnetic field (i.e. $\underline{B} = B\underline{b}$).
The acceleration, to lowest order, is then

$$\frac{d\underline{u}}{dt} = \underline{b}\,\frac{du_\parallel}{dt} + u_\parallel\,\frac{d\underline{b}}{dt}$$

$$= \underline{b}\,\frac{du_\parallel}{dt} + u_\parallel\left[\frac{\partial \underline{b}}{\partial t} + u_\parallel\,\underline{b}.\nabla\underline{b}\right]$$

$$= \underline{b}\,\frac{du_\parallel}{dt} + u_\parallel^2\,\underline{b}.\nabla\underline{b}, \tag{2.10}$$

since $\dfrac{\partial \underline{b}}{\partial t} = 0$ for magnetostatic fields.

We are now in a position to decompose (2.9) into components.
Taking scalar and vector products of $\underline{b}$ with (2.9), and using (2.10),
we obtain

$$m\,\underline{b}.\,\frac{d\underline{u}}{dt} = -\,\mu(\underline{b}.\nabla)B$$

$$\text{i.e. } m\,\frac{du_\parallel}{dt} = -\frac{\mu}{B}\,(\underline{B}.\nabla)B \tag{2.11}$$

and

$$m\underline{b}\times\frac{d\underline{u}}{dt} = e\,\underline{b}\times(\underline{u}\times\underline{b}) - \mu\underline{b}\times\nabla B. \tag{2.12}$$

In (2.12), note that $\underline{b}\times(\underline{u}\times\underline{b}) = \underline{u}_\perp$ and that

$$\underline{b}\times\frac{d\underline{u}}{dt} = u_\parallel\,\underline{b}\times(\underline{b}.\nabla)\underline{b}$$

$$= \frac{u_\parallel}{B^3}\,\underline{B}\times(\underline{B}.\nabla)\underline{B}.$$

This enables us to obtain an expression for $\underline{u}_\perp$

$$\underline{u}_\perp = \frac{\mu}{eB^2}\,\underline{B}\times\nabla B + \frac{mu_\parallel^2}{eB^4}\,\underline{B}\times(\underline{B}.\nabla)\underline{B}. \tag{2.13}$$

In (2.13), the first term is the drift due to shear in the magnetic
field, that is, where there are gradients in the magnetic field
perpendicular to the direction of the field. This term is
proportional to μ, and therefore to w^2. The second term in the
equation is due to curvature of the field lines. We can think of
a particle as being subject to centrifugal acceleration $u_\parallel^2/L$,
perpendicular to $\underline{B}$, L being the radius of curvature. The resultant drift motion is perpendicular both to $\underline{B}$ and to this centrifugal
acceleration.

There is a further simplification if the magnetic field is
source-free, so that $\nabla\times\underline{B} = 0$. Using the vector identity

$$B\nabla B = (\underline{B}.\nabla)\,\underline{B} + \underline{B}\times(\nabla\times\underline{B})$$

we see that, in the source-free case, $B\nabla B = (\underline{B}.\nabla)\underline{B}$, which results in

$$\underline{u}_{\perp} \;=\; \frac{m}{eB^4} \left(\tfrac{1}{2}w^2 + u_{\parallel}^2\right)\, \underline{B} \times (\underline{B}.\nabla)\underline{B}. \qquad (2.14)$$

Finally, we have omitted any reference, in this section on guiding centre motion, to the effect of an electric field $\underline{E}$, or indeed of any force $\underline{F}$, such as a gravitational force, which is independent of the charge e of the particle. This omission was intentional in order not to obscure the analysis.

The equation of motion (2.1), in the case of an electric field $\underline{E}$ and a non-electric force $\underline{F}$, is

$$m\,\frac{d\underline{v}}{dt} \;=\; e(\underline{E} + \underline{v} \times \underline{B}) + \underline{F}. \qquad (2.15)$$

If $\underline{E}$ and $\underline{F}$ are time independent and have non-zero components parallel to $\underline{B}$ the motion along $\underline{B}$ will be dominated by these forces, $E_{\parallel}$ and $F_{\parallel}$. We shall consider here only the case where $E_{\parallel} = F_{\parallel} = 0$. If then we write

$$\underline{v} \;=\; \underline{v}_E + \underline{v}_F + \underline{v}', $$

where

$$\underline{v}_E \;=\; \underline{E} \times \frac{\underline{B}}{B^2}$$

$$\underline{v}_F \;=\; \underline{F} \times \frac{\underline{B}}{eB^2}, \qquad (2.16)$$

we get an equation for $\underline{v}'$

$$m\,\frac{d\underline{v}'}{dt} \;=\; e\,\underline{v}' \times \underline{B}, \qquad (2.17)$$

which is precisely equation (2.1). This shows that the presence of an electric field perpendicular to the magnetic field induces a drift perpendicular to both $\underline{E}$ and $\underline{B}$ and independent of the sign of the charge. The case of a non-electric force $\underline{F}$ is somewhat similar except that charges of opposite sign drift in opposite directions.

2-4 APPLICATIONS OF ORBIT THEORY

(i) Axisymmetric magnetic trap

We discussed in §2-2 the special case of a magnetic field with axial symmetry. Consider now such a field produced by a straight solenoid with extra end coils such that the magnetic field is approximately uniform in the central region, and equal to B_o, rising to a maximum value B_M at either end. The configuration is illustrated in Fig. 2.1.

There is a possibility of containing plasma between two regions of a high magnetic field, if the field varies sufficiently slowly

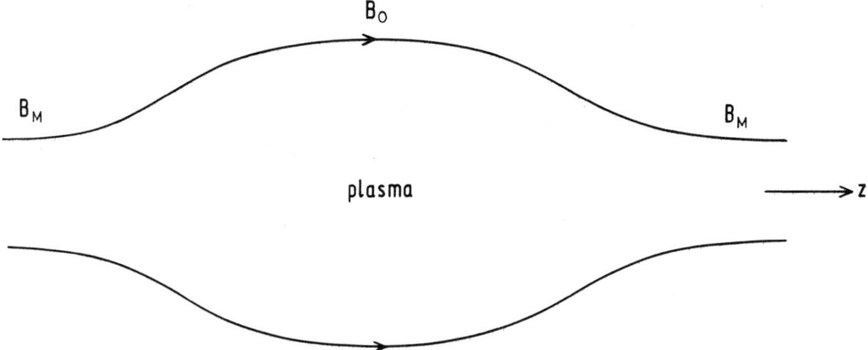

Fig. 2.1 Axisymmetric magnetic trap

that μ is conserved. Consider a charged particle in the centre plane $z = 0$ where $B = B_o$. If, at any z within the trap, the kinetic energy of the particle associated with its motion perpendicular to the field is $W_\perp(z)$, then

$$\frac{W_\perp(z)}{B(z)} = \frac{W_\perp(0)}{B_o} .$$

But $W_\perp(z) + W_\parallel(z) = W$ for all z, $W_\parallel(z)$ and W being the kinetic energy of the motion parallel to the field, and total kinetic energy respectively. Hence, if anywhere $W_\perp(z) = W = W_\perp(0) + W_\parallel(0)$, $W_\parallel(z) = 0$, and the particle is reflected. If θ is the angle between the velocity vector of the particle and the magnetic field at $z = 0$, all particles will be trapped for which $\theta > \theta_o$, where $\sin^2 \theta_o = B_o/B_M = \frac{1}{R}$. R is known as the <u>mirror ratio</u>.

This trapping can only be temporary, since we would expect, in the absence of any loss criterion, that a Maxwellian distribution would become established. The magnetic trap described here is effective only for particles outside the 'loss cone', as shown in Fig. 2.2, and it is expected that interparticle collisions will continually scatter particles into the loss cone. In practice, matters can be much worse, with large-scale instabilities contributing very effectively in filling the loss cone.

It might be thought that such a trap could never be filled in the first place. Several possibilities exist, for example; (a) use of invariance of μ in a magnetic field which increases with time. Then $v_\parallel$ is unaffected but $v_\perp$ can be increased until the trapping criterion is satisfied; (b) a beam of high energy

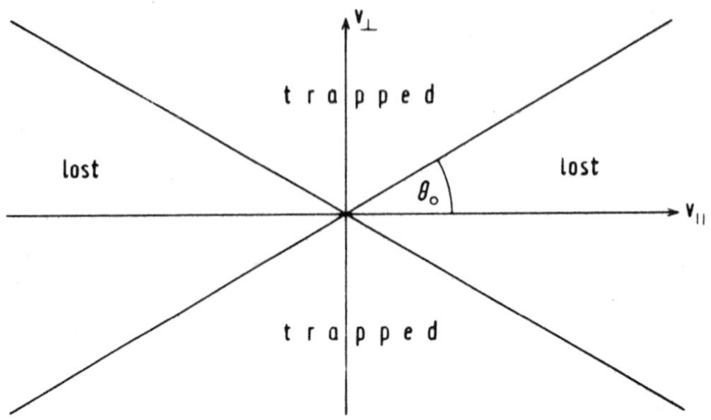

Fig. 2.2 Loss cone

neutral atoms will be unaffected by the magnetic field until ionisation takes place.   This can be achieved either by collision
with a neutral background gas already in the trap, by collision
with plasma already trapped, or by the magnetic field itself.   In
the latter process, known as Lorentz ionisation, the electrons and
ions experience a sufficiently different magnetic force that the
atomic binding can be overcome;   this process is inefficient unless
the magnetic field strength is in excess of 1 tesla.

(ii) Toroidal containment

The current interest in toroidal containment, mainly in Tokamak
devices (cf. Chapter 5) has revealed some interesting applications
of single particle orbit theory.   Suppose we have a configuration
in which the magnetic field is purely toroidal (i.e. no component
in a minor cross section).   A single particle will have no
electric drift, but there will be a drift due to both curvature
and non-zero ∇B.   This drift will be in the z-direction, the axis
of the torus, positive for an ion and negative for an electron, as
shown in Fig. 2.3.

In a low density plasma, ions and electrons drift initially in
opposite directions, producing a charge separation.   The effect of
the resulting electric field $\underline{E}$, acting in the negative z-direction,
is to produce a radially outward drift away from the z-axis.   This
is shown in Fig. 2.4.

A purely toroidal magnetic field is thus unable to contain a low
density plasma.   We shall come to the same conclusion, but for a
different reason, when we discuss the problem again in §3-4, from a
magnetofluid approach.

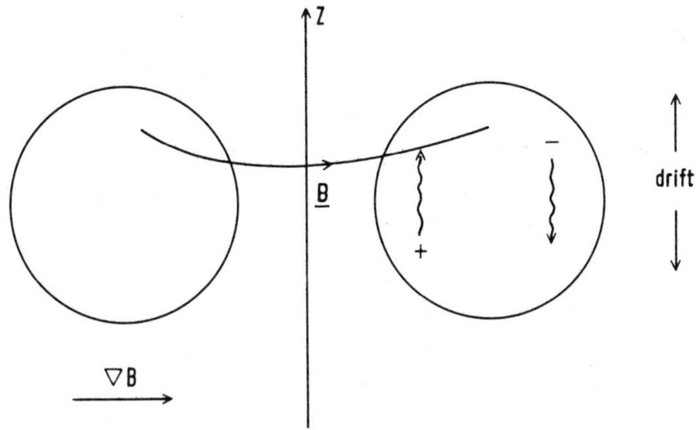

Fig. 2.3 Single particle drift in toroidal magnetic field

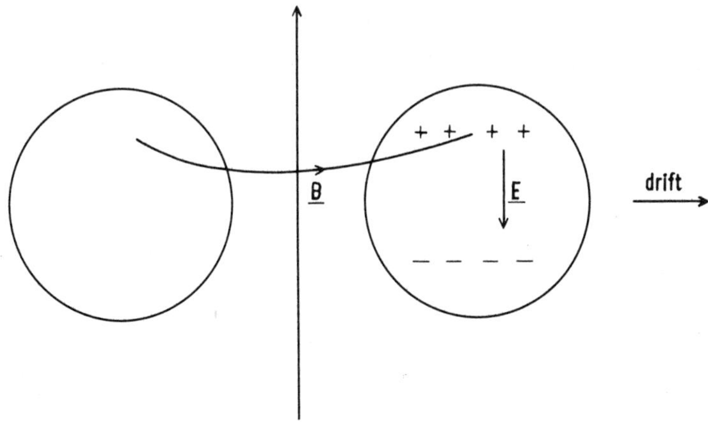

Fig. 2.4 Outward drift due to build up of space charge

The Tokamak configuration relies on toroidal plasma currents to generate a small rotational transform, the result of combining a strong toroidal magnetic field and a relatively weak poloidal field, the latter appearing as a consequence of the plasma currents.   The object is to avoid closure of magnetic field lines after once around the torus (or even an integral number of times). Consequently the magnetic field generates a set of closed nested magnetic surfaces.

In 1968, Galeev and Sagdeev[5] published an important paper in which they calculated the flux of particles across the magnetic field, that is, normal to the magnetic surfaces.   We shall touch only on one aspect of their work which is relevant to the topic

of this chapter. Most particles circulate around the torus, but
a certain class do not. These have a sufficiently small component
of velocity parallel to the magnetic field, which is predominantly
toroidal. It can be shown quite generally that a toroidal magnetic
field, whether produced by a current along the z-axis (cf. Fig.
2.4), or otherwise, is not uniform but is inversely proportional
to r, the radial distance from the z-axis. It is therefore
stronger in regions close to the toroidal axis. A small poloidal
component, typically only a few percent of the toroidal field,
does not substantially change the total B, though it is critical in
defining the topology of the field. Charged particles can then be
reflected between mirror points and be trapped. The effect of
curvature and B drifts is also present, causing particles to map
out so-called drift surfaces. In Fig. 2.5 the projection of a
typical drift surface onto a minor cross section is shown. Its
shape has earned it the name 'banana orbit'.

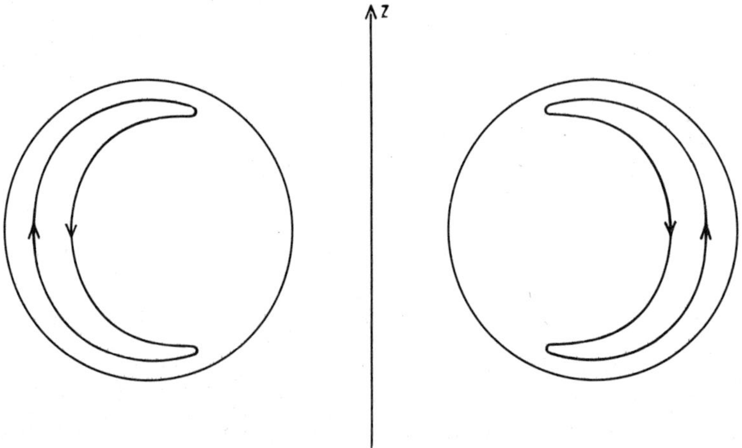

Fig. 2.5 Projection onto minor cross section of trapped
particle orbit

26

Chapter 3

FLUID THEORY

In Chapter 1, we saw that in general a high temperature plasma has a low electrical resistivity. We also assumed in some of the derivations of plasma properties that we could treat the plasma as though it were a fluid. In Chapter 2, we saw that the presence of a magnetic field greatly complicates the motion of a single charged particle, and some of these effects show up in a rigorous treatment based on kinetic theory. The aim of this chapter is to explore the main consequences of adopting a fluid picture, without at this stage questioning its physical validity. This will allow us to begin the study of the interaction between a plasma and a magnetic field.

In much of what follows, we shall neglect transport processes. Thus, we shall suppose electrical resistivity η to be zero. This will be a reasonable approximation for processes with a time-scale short compared with that required for a magnetic field to diffuse through the plasma. Similar arguments can be given in respect of the neglect of thermal conductivity and viscosity.

3-1 CONSERVATION OF MAGNETIC FLUX

In a plasma with zero electrical resistivity, the magnetic flux Φ through any closed contour in the plasma, each element of which moves with the local velocity $\underline{v}$ of the plasma, remains constant.

$$\Phi = \int_A \underline{B}.d\underline{A}.$$

The change in Φ through a moving surface A bounded by a contour C is made up of two parts:

(i) Change of $\underline{B}$ with time at various points of A

$$\left(\frac{\partial \Phi}{\partial t}\right)_1 = \int_A \frac{\partial \underline{B}}{\partial t} . d\underline{A}.$$

(ii) Motion of contour C

If $d\underline{l}$ is an element of C, $\underline{v} \times d\underline{l}$ is the area swept out by $d\underline{l}$ per

unit time. Hence the flux crossed per unit time is $\underline{B}.\underline{v} \times d\underline{l}$.
Hence

$$\left(\frac{\partial \Phi}{\partial t}\right)_2 = \int_C \underline{B} \cdot \underline{v} \times d\underline{l} = \int_C \underline{B} \times \underline{v} \cdot d\underline{l}$$

$$= \int_A \nabla \times (\underline{B} \times \underline{v}) \cdot d\underline{A}, \quad \text{by Stokes' theorem.}$$

Using Maxwell's equation, we can write

$$\left(\frac{\partial \Phi}{\partial t}\right)_1 = -\int_A \nabla \times \underline{E} \cdot d\underline{A}.$$

Hence the total rate of change of Φ is

$$\frac{d\Phi}{dt} = -\int_A \nabla \times (\underline{E} + \underline{v} \times \underline{B}) \cdot d\underline{A}.$$

The condition that Φ remains constant for any arbitrary contour
moving with the plasma is thus

$$\nabla \times (\underline{E} + \underline{v} \times \underline{B}) = 0 \tag{3.1}$$

at every point in the plasma.

Now, in our simple model we have assumed zero resistivity.
Since the relation between $\underline{E}$ and $\underline{j}$ is usually written $\underline{E} = \eta\underline{j}$, we
would therefore suppose $\underline{E} = 0$ for consistency. However, $\underline{E}$ and $\underline{B}$
are measured relative to a stationary observer, not one moving with
every plasma element. What is appropriate is that the <u>local</u>
electric field $\underline{E}'$, in the reference frame of the plasma element,
should vanish. Since $\underline{E}' = \underline{E} + \underline{v} \times \underline{B}$, our condition $\eta = 0$ implies
$\underline{E} + \underline{v} \times \underline{B} = 0$, which in turn results in the conservation of Φ. It
is loosely stated that the plasma is <u>tied</u> to the magnetic field,
or conversely that the field is <u>frozen</u> into the plasma. The meaning is clear in terms of magnetic flux conservation.

3-2 FINITE RESISTIVITY η

We showed in §3-1 that in the case of $\eta = 0$ no motion of the plasma
is possible normal to the magnetic field without a compensating
electric field $\underline{E} = -\underline{v} \times \underline{B}$.

For finite resistivity, this result no longer holds. We assume
that it is replaced by a simple Ohm's Law

$$\underline{E} + \underline{v} \times \underline{B} = \eta\underline{j}. \tag{3.2}$$

$$\text{Then } \nabla \times (\underline{E} + \underline{v} \times \underline{B}) = \eta\nabla \times \underline{j}$$

taking η to be constant.

Thus,
$$-\frac{\partial \underline{B}}{\partial t} + \nabla \times (\underline{v} \times \underline{B}) = \frac{\eta}{\mu_o} \nabla \times \nabla \times \underline{B},$$

which may finally be written as

$$\frac{\partial \underline{B}}{\partial t} - \nabla \times (v \times B) = \frac{\eta}{\mu_o} \nabla^2 \underline{B}. \tag{3.3}$$

This is an equation for the motion of the magnetic field lines. Using our earlier results for the magnetic flux Φ, we obtain on integrating over any arbitrary surface moving with the plasma

$$\frac{d\Phi}{dt} = \frac{\eta}{\mu_o} \nabla^2 \Phi, \tag{3.4}$$

which is a diffusion equation for Φ, with diffusion coefficient $\frac{\eta}{\mu_o}$.

3-3 MAGNETIC PRESSURE

A magnetic field can exert a stress on a plasma, which in simple situations is equivalent to a pressure. We first obtain the equations of motion for a plasma in an electric and magnetic field. We could proceed by considering the equations for a conducting fluid, but it is useful to take, as our starting point, a plasma consisting of a fluid of two components, an electron and an ion fluid, interacting with each other.

Consider a small element of this fluid, in which the electron component has density n_e, velocity $\underline{v}_e$, and pressure p_e. The ion component has, similarly, n_i, $\underline{v}_i$, and p_i. We now write down the equations of motion for ions and electrons separately:

$$n_i\, m_i\, \frac{d\underline{v}_i}{dt} = -\nabla p_i + n_i e\, (\underline{E} + \underline{v}_i \times \underline{B}) \tag{3.5}$$

$$n_e\, m_e\, \frac{d\underline{v}_e}{dt} = -\nabla p_e - n_e e\, (\underline{E} + \underline{v}_e \times \underline{B}). \tag{3.6}$$

In obtaining these equations, we have made an important assumption, that ions and electrons interact only through $\underline{E}$ and $\underline{B}$. The latter may of course arise in part from current and charge densities in the plasma itself. What we are neglecting is the effect of particle collisions, which would be manifest in the equations of motion as a viscous term, and therefore not in the spirit of the physical approximations being made in ideal magnetofluid theory.

The volume element chosen is necessarily small compared with the dimensions of the system. We shall assume that it is nevertheless large compared with the screening length λ. In such a case, we expect all charge fluctuations to average out to zero, yielding

$n_i \simeq n_e = n$. This is usually termed the approximation of charge quasi-neutrality. We now add the ion and electron equations to give

$$\rho \frac{d\underline{v}}{dt} = -\nabla p + \underline{j} \times \underline{B}, \qquad (3.7)$$

where

$$\rho \underline{v} = n (m_i \underline{v}_i + m_e \underline{v}_e),$$

so that ρ is the mass density of the plasma fluid, and $\underline{v}$ its velocity.

$$p = p_i + p_e = \text{total plasma pressure}$$

$$\underline{j} = n_i e \underline{v}_i - n_e e \underline{v}_e = \text{net current density.}$$

Hence, we note that $\underline{E}$ takes no explicit part and we have reduced the description of the plasma to that of a one-component conducting fluid.

A simpler equation still is obtained if the plasma is in static equilibrium. Then

$$\nabla p = \underline{j} \times \underline{B}. \qquad (3.8)$$

In order to define the steady state completely we need also Maxwell's equations for $\underline{j}$ and $\underline{B}$ for steady currents and fields:

$$\nabla \times \underline{B} = \mu_o \underline{j} \qquad (3.9)$$

$$\nabla . \underline{B} = 0. \qquad (3.10)$$

It is possible now to see how a magnetic pressure can arise. Suppose that $\underline{B}$ is non-uniform, but everywhere parallel to the x-axis, $\underline{B} = B\hat{\underline{x}}$. This implies that

$$(\underline{B}.\nabla) \underline{B} = B \frac{\partial B}{\partial x} \hat{\underline{x}}.$$

But $\nabla . \underline{B} = 0$ implies that $\frac{\partial B}{\partial x} = 0$, so $(\underline{B}.\nabla) \underline{B} = 0$. Eliminating $\underline{j}$ from equations (3.8) and (3.9) we obtain

$$\nabla p = \frac{1}{\mu_o} (\nabla \times \underline{B}) \times \underline{B}$$

$$= \frac{1}{\mu_o} \{(\underline{B}.\nabla) \underline{B} - \nabla (\tfrac{1}{2}\underline{B}^2)\}.$$

Hence, $\qquad \nabla \left(p + \frac{B^2}{2\mu_o}\right) = 0, \qquad (3.11)$

which means that $p + \frac{B^2}{2\mu_o} = $ constant. Thus pressure falls where the magnitude of the field increases. In fact, just as p may be interpreted as the energy density of the fluid, so $\frac{B^2}{2\mu_o}$ is the magnetic field energy density. The result above therefore establishes the concept of magnetic pressure, and is the basis of magnetic plasma confinement.

3-4 PLASMA EQUILIBRIUM

In this section we give a brief description of some cylindrical and toroidal plasma equilibria under magnetic confinement.

(i) Axial pinch

If a current $\underline{j}$ flows axially in a cylindrical plasma, an azimuthal magnetic field $\underline{B}$ is generated, and the force $\underline{j} \times \underline{B}$ acts radially inwards. This is known as the pinch effect, which for a sufficiently large current density may balance the outward force ∇p of the plasma. The equilibrium equation

$$\nabla p - \underline{j} \times \underline{B} = 0$$

becomes, in the case of axial symmetry,

$$\frac{dp}{dr} + \frac{B_\theta}{\mu_o r} \frac{d(B_\theta r)}{dr} = 0, \tag{3.12}$$

which may be written in the form

$$\frac{d}{dr} \left(p + \frac{B_\theta^2}{2\mu_o} \right) + \frac{B_\theta^2}{\mu_o r} = 0. \tag{3.13}$$

Equation (3.13) shows that, as well as a magnetic pressure similar to that in (3.11), there is a term $B_\theta^2/\mu_o r$ which arises from the curvature of the magnetic field.

(ii) Theta pinch

In this second illustration of magnetic confinement a large current j_θ is caused to flow azimuthally around the plasma cylinder, generating an axial magnetic field B_z, and again causing the plasma to pinch. In this case, the field lines are straight, and the equilibrium condition is just

$$p + \frac{B_z^2}{2\mu_o} = \text{constant.} \tag{3.14}$$

Experimentally, the axial pinch is set up by discharging a large current through the plasma by means of electrodes placed at the extremities of the cylinder. In the case of the theta pinch, a copper strap encircles the cylinder except for a small gap where the current is led in and out.

It is possible to combine these two effects into the most general equilibrium configuration with cylindrical symmetry. Then

$\underline{B} = (0, B_\theta(r), B_z(r))$, since there can be no radial component to preserve $\nabla.\underline{B} = 0$.   The equilibrium condition is then a combination of (3.13) and (3.14):

$$\frac{d}{dr}\left[p + \frac{B_\theta^2 + B_z^2}{2\mu_o}\right] + \frac{B_\theta^2}{\mu_o r} = 0. \qquad (3.15)$$

This configuration has been widely studied in investigations of plasma stability.   We shall come back to this point later in the chapter.

(iii) Toroidal equilibria

End effects lead to many difficulties in attempting to set up and maintain either an axial or theta pinch.   One way of avoiding electrodes, and end losses, with subsequent cooling of the plasma and loss of confinement, is to confine the plasma in a torus rather than a cylinder.   The plasma loop can then be considered as the secondary in a transformer, and a large current induced around the torus.

We can easily show that there is no toroidal counterpart to the theta pinch.   That is, no equilibrium is possible if the magnetic field is purely toroidal,

$$\underline{B} = B(r, z)\ \hat{\underline{\phi}}.$$

(The co-ordinate system adopted is shown in Fig. 3.1.)

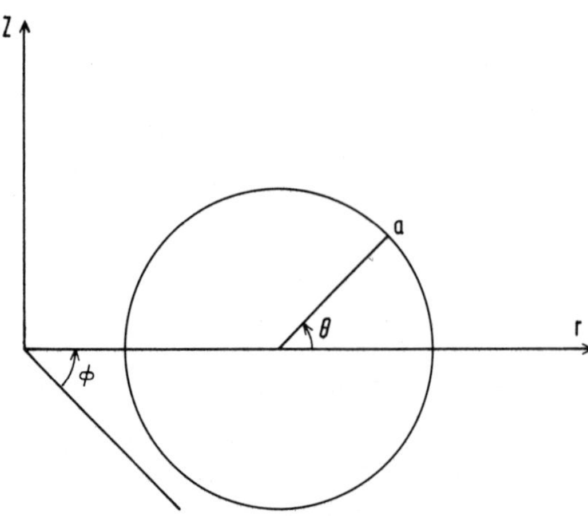

Fig. 3.1 Co-ordinates (r, ϕ, z) for torus

We now write down the equilibrium equations in components, noting

that $\nabla.\underline{B} = 0$ is already satisfied, and assuming that p also is
independent of ϕ.   Then,

$$\mu_o \frac{\partial p}{\partial r} = - \frac{B}{r} \frac{\partial}{\partial r} (rB) = - \frac{\partial}{\partial r} (\tfrac{1}{2}B^2) - \frac{B^2}{r} \qquad (3.16)$$

$$\mu_o \frac{\partial p}{\partial z} = - \frac{B}{r} \frac{\partial}{\partial z} (rB) = - \frac{\partial}{\partial z} (\tfrac{1}{2}B^2), \qquad (3.17)$$

which may be rewritten in the form

$$\frac{\partial}{\partial r} (\mu_o p + \tfrac{1}{2}B^2) = - \frac{B^2}{r} \qquad (3.18)$$

$$\frac{\partial}{\partial z} (\mu_o p + \tfrac{1}{2}B^2) = 0. \qquad (3.19)$$

From (3.19), we have that $\mu_o p + \tfrac{1}{2}B^2$ is independent of z, so that,
from (3.18), B is separately independent of z and hence likewise
is p.   But if $p = p(r)$, containment is impossible.   For example,
on the straight line through the centre of the torus parallel to
the z-axis, r is constant, therefore $p(r)$ is constant.   The
pressure is therefore the same at the circumference of the torus
as at the centre.

Thus, no toroidal equivalent to the theta pinch exists.   We
now show that other equilibria may exist, but that to find one
analytically poses a severe mathematical problem.   We shall still
restrict ourselves to toroidal symmetry, but now all three
components of magnetic field are present, $\underline{B} = (B_r, B_\phi, B_z)$.   The
condition $\nabla.\underline{B} = 0$ is automatically satisfied, if we define B_r and
B_z in terms of a stream function $\psi(r, z)$:

$$B_r = - \frac{1}{r} \frac{\partial \psi}{\partial z} ; \qquad B_z = \frac{1}{r} \frac{\partial \psi}{\partial r} . \qquad (3.20)$$

$B_\phi (r, z)$ is at this stage quite arbitrary, as also is $p(r, z)$.

The components of the equilibrium equation $\nabla p - \underline{j} \times \underline{B} = 0$ now
become

$$\frac{\partial p}{\partial r} + \frac{1}{r^2} \frac{\partial \psi}{\partial r} \Delta^2 \psi + \frac{B_\phi}{r} \frac{\partial}{\partial r} (rB_\phi) = 0 \qquad (3.21)$$

$$\frac{1}{r^2} \frac{\partial \psi}{\partial r} \frac{\partial}{\partial z} (rB_\phi) - \frac{1}{r^2} \frac{\partial \psi}{\partial z} \frac{\partial}{\partial r} (rB_\phi) = 0 \qquad (3.22)$$

$$\frac{\partial p}{\partial z} + B_\phi \frac{\partial B_\phi}{\partial z} + \frac{1}{r^2} \frac{\partial \psi}{\partial z} \Delta^2 \psi = 0, \qquad (3.23)$$

where

$$\Delta^2 \psi = \frac{\partial^2 \psi}{\partial r^2} + \frac{\partial^2 \psi}{\partial z^2} - \frac{1}{r} \frac{\partial \psi}{\partial r} .$$

Equation (3.22), when written as the Jacobian

$$\frac{\partial(\psi,\ rB_\phi)}{\partial(r,z)} = 0 \qquad\qquad (3.24)$$

implies that rB_ϕ is a function of r and z only as a functional of ψ:

$$rB_\phi \equiv f(\psi). \qquad\qquad (3.25)$$

Equations (3.21) and (3.23) are then, respectively,

$$\frac{\partial p}{\partial r} + \frac{1}{r^2}\frac{\partial\psi}{\partial r}\left[\Delta^2\psi + f\frac{df}{d\psi}\right] = 0$$

$$\frac{\partial p}{\partial z} + \frac{1}{r^2}\frac{\partial\psi}{\partial z}\left[\Delta^2\psi + f\frac{df}{d\psi}\right] = 0,$$

from which we can easily show that

$$\frac{\partial(\psi,\ p)}{\partial(r,\ z)} = 0, \qquad\qquad (3.26)$$

so that $p \equiv g(\psi)$.

We can now write down an equation for ψ

$$\Delta^2\psi + f\frac{df}{d\psi} + r^2\frac{dg}{d\psi} = 0, \qquad\qquad (3.27)$$

where f and g are arbitrary functionals of ψ. Surfaces of constant ψ can be identified with magnetic surfaces, on which the pressure is also constant. This last result is most easily obtained from the basic equilibrium equation $\nabla p = \underline{j} \times \underline{B}$, which implies that

$$\underline{B}.\nabla p = 0 \qquad\qquad (3.28)$$

$$\underline{j}.\nabla p = 0. \qquad\qquad (3.29)$$

(3.28) shows that ∇p is everywhere perpendicular to $\underline{B}$, so that if $\underline{B}$ - lines generate a surface, p will be constant on such a surface. A similar result can be obtained from (3.29), involving currents.

The question raised here about the existence of magnetic surfaces is very important. We observe that the absence of toroidal equilibria in the case of a purely toroidal magnetic field, investigated above, is also a case where magnetic surfaces do not exist.

In conclusion, to obtain a general toroidal equilibrium, it is necessary to obtain a solution of equation (3.27), given some arbitrary $f(\psi)$ and $g(\psi)$. However, only after ψ is known can we know B_ϕ and p as functions of r and z. (cf. Thompson in Bibliography).

3-5 PROPAGATION OF WAVES
(i) Incompressible plasma
An incompressible plasma, in which the density remains constant,

$\rho = \rho_o$, behaves quite differently from an incompressible non-conducting fluid. In the latter, waves cannot propagate, for we can show that longitudinal disturbances, which involve compressions and rarefactions, are propagated instantaneously, while transverse waves are prohibited if the forces acting on the fluid are derivable from a potential. To derive these results, we note that the equation of continuity becomes $\nabla \cdot \underline{v} = 0$, so that in wave motion with wave number $\underline{k}$ and frequency ω, $\underline{v} = \underline{v}_o \exp(i\underline{k}\cdot\underline{r} - i\omega t)$, we have $\underline{k}\cdot\underline{v} = 0$. Thus $\underline{k}$ and $\underline{v}$ are necessarily at right angles, and so only transverse waves are possible. However, the assumption of conservative forces, implying a potential ϕ and an equation of motion

$$\rho_o \frac{d\underline{v}}{dt} = -\nabla(p + \phi) \tag{3.30}$$

implies that $\qquad \nabla \times \dfrac{d\underline{v}}{dt} = 0.$

Since $\qquad \dfrac{d\underline{v}}{dt} = \dfrac{\partial \underline{v}}{\partial t} + (\underline{v}\cdot\nabla)\underline{v} = -i\omega\underline{v} + i(\underline{k}\cdot\underline{v})\underline{v}$

and using the result $\underline{k}\cdot\underline{v} = 0$, we get

$$\omega\underline{k} \times \underline{v} = 0. \tag{3.31}$$

Since $\underline{k} \times \underline{v} \neq 0$, (3.31) yields $\omega = 0$ and the disturbance cannot propagate.

In a magnetised plasma, this result no longer holds, and transverse waves are possible.

Again, we consider the simplest physical situation of a uniform incompressible perfectly conducting fluid at rest in a uniform magnetic field $\underline{B}_o$. Let the system be slightly disturbed by introduction of a small velocity disturbance $\underline{v}_1$, giving rise to other small perturbations $\underline{B}_1$ and p_1 in the magnetic field and pressure, respectively.

Our starting point is the set of equations describing the plasma motion:

$$\rho_o \frac{d\underline{v}}{dt} = -\nabla p + \frac{1}{\mu_o} (\nabla \times \underline{B}) \times \underline{B} \tag{3.32}$$

$$\frac{\partial \underline{B}}{\partial t} = \nabla \times (\underline{v} \times \underline{B}) \tag{3.33}$$

$$\nabla \cdot \underline{B} = 0 \tag{3.34}$$

$$\nabla \cdot \underline{v} = 0. \tag{3.35}$$

We shall retain only first order quantities, so that

$$\frac{d}{dt} \equiv \frac{\partial}{\partial t} + \underline{v}\cdot\nabla = \frac{\partial}{\partial t}.$$

Then, equations (3.32) and (3.33), to first order, yield

$$\rho_o \frac{\partial \underline{v}_1}{\partial t} = -\nabla p_1 + \frac{1}{\mu_o} (\nabla \times \underline{B}_1) \times \underline{B}_o \qquad (3.36)$$

$$\frac{\partial \underline{B}_1}{\partial t} = \nabla \times (\underline{v}_1 \times \underline{B}_o). \qquad (3.37)$$

Let $\underline{\psi}_1 = \nabla \times \underline{v}_1$ = fluid vorticity, apply to (3.36) the operation $\frac{\partial}{\partial t} \nabla \times$ and $\nabla \times$ to (3.37). We obtain an equation for $\underline{\psi}_1$ from the resulting equations, thus:

$$\rho_o \frac{\partial^2 \underline{\psi}_1}{\partial t^2} = \frac{1}{\mu_o} \frac{\partial}{\partial t} \nabla \times \left[(\nabla \times \underline{B}_1) \times \underline{B}_o \right]$$

$$= \frac{1}{\mu_o} (\underline{B}_o \cdot \nabla) \nabla \times \frac{\partial \underline{B}_1}{\partial t} \qquad (3.38)$$

$$\nabla \times \frac{\partial \underline{B}_1}{\partial t} = \nabla \times \nabla \times (\underline{v}_1 \times \underline{B}_o)$$

$$= \nabla \times \left[(\underline{B}_o \cdot \nabla) \underline{v}_1 \right], \qquad (3.39)$$

so that

$$\rho_o \frac{\partial^2 \underline{\psi}_1}{\partial t^2} = \frac{1}{\mu_o} (\underline{B}_o \cdot \nabla)^2 \underline{\psi}_1, \qquad (3.40)$$

which is in the form of a wave equation.

Again we suppose that $\underline{\psi}_1 \propto \exp(i\underline{k}.\underline{r} - i\omega t)$.

Defining $V = \frac{\omega}{|\underline{k}|}$ = phase velocity of wave and θ = angle between $\underline{B}_o$ and the direction of propagation, we obtain

$$V^2 = \frac{B_o^2}{\mu_o \rho_o} \cos^2 \theta. \qquad (3.41)$$

Note that a fundamental velocity $C_A = (B_o^2/\mu_o \rho_o)^{\frac{1}{2}}$ appears in (3.41). This is termed the <u>Alfven</u> velocity. The waves are of course transverse in nature, since $\underline{k}.\underline{v}_1 = 0$. Their velocity in any particular direction making an angle θ to the magnetic field is

$$V_A = C_A \cos \theta, \qquad (3.42)$$

which is conveniently shown on a polar diagram, Fig. 3.2.

It is instructive to compare this situation with waves in a string, which propagate with velocity $C = (T/\rho_o)^{\frac{1}{2}}$, where T is the tension and ρ_o the linear density, in appropriate units. We have seen in §3-1 that a magnetic field and plasma are "tied", so that it may be possible to consider a field line to be "loaded" with plasma of density ρ_o. Further, in terms of the Maxwell stress

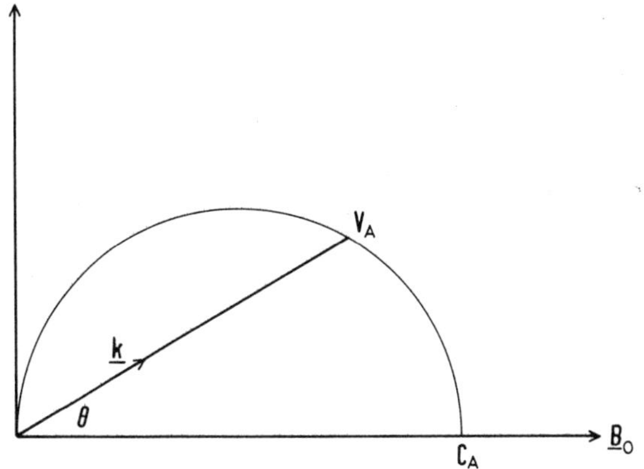

Fig. 3.2 Polar diagram for velocity of propagation of
incompressible waves.

tensor, the line is under a tension B_o^2/μ_o, so that $C = (B_o^2/\mu_o\rho_o)^{\frac{1}{2}}$
$= C_A$. We thus obtain the correct result for waves propagating
<u>along</u> the magnetic field.

(ii) Compressible plasma

This case is more complicated, in particular by the full equation
of continuity. We shall assume also that compressions are
sufficiently rapid, as in sound waves, that no heat is exchanged
in the process, and isentropic flow results. Our starting set of
equations is thus

$$\rho \frac{d\underline{v}}{dt} = -\nabla p + \frac{1}{\mu_o} (\nabla \times \underline{B}) \times \underline{B} \tag{3.43}$$

$$\frac{\partial \underline{B}}{\partial t} = \nabla \times (\underline{v} \times \underline{B}) \tag{3.44}$$

$$\frac{\partial \rho}{\partial t} + \nabla.(\rho\underline{v}) = 0 \tag{3.45}$$

$$\frac{d}{dt} (p\rho^{-\gamma}) = 0 \tag{3.46}$$

$$\nabla.\underline{B} = 0. \tag{3.47}$$

In (3.46), $\gamma =$ ratio of specific heats C_p/C_v. We consider the same
physical situation as in (i), that is, a uniform plasma with p_o,
ρ_o, $\underline{B}_o$. On this we superimpose small fluctuations p_1, ρ_1, $\underline{B}_1$
resulting from $\underline{v}_1$, all varying as $\exp(i\underline{k}.\underline{r} - i\omega t)$. The linearised
set of equations is

$$\rho_o \frac{\partial \underline{v}_1}{\partial t} = -\nabla p_1 + \frac{1}{\mu_o} (\nabla \times \underline{B}_1) \times \underline{B}_o \tag{3.48}$$

37

$$\frac{\partial \underline{B}_1}{\partial t} = \nabla \times (\underline{v}_1 \times \underline{B}_o) \tag{3.49}$$

$$\frac{\partial \rho_1}{\partial t} + \rho_o \, \nabla \cdot \underline{v}_1 = 0 \tag{3.50}$$

$$\frac{\partial}{\partial t} \left(p_1 - \frac{\gamma p_o}{\rho_o} \rho_1 \right) = 0 \tag{3.51}$$

$$\nabla \cdot \underline{B}_1 = 0, \tag{3.52}$$

which reduce to the following set of algebraic equations

$$\rho_o \, \omega \underline{v}_1 = -\underline{k}p_1 + (\underline{k} \times \underline{B}_1) \times \frac{\underline{B}_o}{\mu_o} \tag{3.53}$$

$$\omega \underline{B}_1 = \underline{k} \times (\underline{v}_1 \times \underline{B}_o) \tag{3.54}$$

$$\omega \rho_1 + \rho_o \, \underline{k} \cdot \underline{v}_1 = 0 \tag{3.55}$$

$$\omega \left(p_1 - \frac{\gamma p_o}{\rho_o} \rho_1 \right) = 0 \tag{3.56}$$

$$\underline{k} \cdot \underline{B}_1 = 0. \tag{3.57}$$

(3.53) and (3.54) can be simplified further as

$$\rho_o \, \omega \underline{v}_1 = -\underline{k} \left(p_1 + \frac{\underline{B}_o \cdot \underline{B}_1}{\mu_o} \right) + (\underline{k} \cdot \underline{B}_o) \frac{\underline{B}_1}{\mu_o} \tag{3.58}$$

$$\omega \underline{B}_1 = (\underline{k} \cdot \underline{B}_o) \, \underline{v}_1 - (\underline{k} \cdot \underline{v}_1) \underline{B}_o. \tag{3.59}$$

The object again is to find all possible solutions of the set of equations (3.55) - (3.59) in the form of an equation for $V = \frac{\omega}{k}$.

From (3.56), either $\omega = 0$ or $p_1 = \frac{\gamma p_o}{\rho_o} \rho_1$.

We therefore assume $\omega \neq 0$ and proceed to eliminate p_1, ρ_1 and $\underline{B}_1$ to obtain an equation for $\underline{v}_1$:

$$p_1 = \frac{\gamma p_o}{\rho_o} \rho_1$$

$$\rho_1 = -\frac{\rho_o}{\omega} \underline{k} \cdot \underline{v}_1 \tag{3.60}$$

$$\underline{B}_1 = \frac{1}{\omega} \{ (\underline{k} \cdot \underline{B}_o) \, \underline{v}_1 - (\underline{k} \cdot \underline{v}_1) \, \underline{B}_o \}$$

Substituting (3.60) into (3.58) we obtain, after some rearrangement,

$$\left[\omega^2 - \frac{(\underline{k} \cdot \underline{B}_o)^2}{\rho_o \mu_o} \right] \underline{v}_1$$

$$= \underline{k} \cdot \underline{v}_1 \left[\underline{k} \left(\frac{\gamma p_o}{\rho_o} + \frac{B_o^2}{\rho_o \mu_o} \right) - \underline{B}_o \frac{\underline{k} \cdot \underline{B}_o}{\mu_o \rho_o} \right] - \underline{k} \frac{(\underline{B}_o \cdot \underline{v}_1)(\underline{B}_o \cdot \underline{k})}{\mu_o \rho_o} . \tag{3.61}$$

Let $\underline{b}$, $\underline{z}$ be unit vectors along $\underline{B}_o$, $\underline{k}$, $C_A = (B_o^2/\mu_o \rho_o)^{\frac{1}{2}}$, $C_S = (\gamma p_o/\rho_o)^{\frac{1}{2}}$ = sound velocity. Then, (3.61) may be written as an equation for the phase velocity V,

$$\left[V^2 - C_A^2 (\underline{b} \cdot \underline{z})^2 \right] \underline{v}_1 = (\underline{z} \cdot \underline{v}_1) \left[\underline{z}(C_A^2 + C_S^2) - \underline{b} \, C_A^2 (\underline{b} \cdot \underline{z}) \right] -$$

$$- C_A^2 (\underline{b} \cdot \underline{v}_1)(\underline{b} \cdot \underline{z})\underline{z} . \tag{3.62}$$

To obtain a solution for V, we first look at the component perpendicular to $\underline{z}$

$$\left[V^2 - C_A^2 (\underline{b} \cdot \underline{z})^2 \right] \underline{z} \times \underline{v}_1 = C_A^2 (\underline{z} \cdot \underline{v}_1)(\underline{b} \cdot \underline{z})(\underline{b} \times \underline{z}) . \tag{3.63}$$

If now $\underline{z} \cdot \underline{v}_1 = 0$, implying $\underline{k} \cdot \underline{v}_1 = 0$ or $\nabla \cdot \underline{v}_1 = 0$, the RHS of (3.63) vanishes and consequently the square bracket on the LHS must also vanish, which is just the condition for Alfven waves again, so

$$V = \pm C_A \cos\theta, \quad \text{where } \cos\theta = \underline{b} \cdot \underline{z}.$$

For these waves, (3.62) shows that $\underline{b} \cdot \underline{v}_1 = 0$, that is, $\underline{v}_1$ is normal to both $\underline{B}_o$ and $\underline{k}$.

Since (3.62) is in effect a set of three scalar equations for V^2, we should expect to obtain two more roots for V^2, which we now proceed to find.

So far, we have investigated perturbations normal to the plane defined by $\underline{B}_o$, $\underline{k}$. We now look at perturbations $\underline{v}_1$ in that plane, which can be done most easily by taking components of (3.62) parallel to $\underline{z}$ and $\underline{b}$. These are, respectively,

$$(V^2 - C_A^2 - C_S^2)(\underline{z} \cdot \underline{v}_1) + C_A^2 (\underline{b} \cdot \underline{z})(\underline{b} \cdot \underline{v}_1) = 0 \tag{3.64}$$

$$V^2 \, \underline{b} \cdot \underline{v}_1 - C_S^2 (\underline{z} \cdot \underline{v}_1)(\underline{b} \cdot \underline{z}) = 0 . \tag{3.65}$$

Eliminating $\underline{b} \cdot \underline{v}_1$ from (3.64) and (3.65), we find

$$V^2 (V^2 - C_A^2 - C_S^2)(\underline{z} \cdot \underline{v}_1) + C_A^2 C_S^2 (\underline{b} \cdot \underline{z})^2 (\underline{z} \cdot \underline{v}_1) = 0,$$

which, since $\underline{z} \cdot \underline{v}_1 \neq 0$, reduces to

$$V^2 (V^2 - C_A^2 - C_S^2) + C_A^2 C_S^2 \cos^2\theta = 0. \qquad (3.66)$$

Hence, the four remaining roots for V are

$$V = \pm \left\{ \tfrac{1}{2} \left[C_S^2 + C_A^2 \pm \left((C_S^2 + C_A^2)^2 - 4C_S^2 C_A^2 \cos^2\theta \right)^{\frac{1}{2}} \right] \right\}^{\frac{1}{2}}.$$

This analysis shows that, in addition to Alfven waves, there exist two other types of wave involving compressions of the plasma. These are known as the fast and slow magnetosonic waves V_+, V_-. We now discuss some of their properties.

(a) Propagation perpendicular to $\underline{B}_o$, $\theta = \dfrac{\pi}{2}$.

$$V_+ = (C_S^2 + C_A^2)^{\frac{1}{2}}$$

$$V_- = 0.$$

(b) Propagation parallel to $\underline{B}_o$, $\theta = 0$.

$$V_+ = \max (C_S, C_A)$$

$$V_- = \min (C_S, C_A).$$

(c) Propagation at arbitrary angle θ to $\underline{B}_o$

V_+, V_- are shown in the polar diagram, Fig. 3.3.

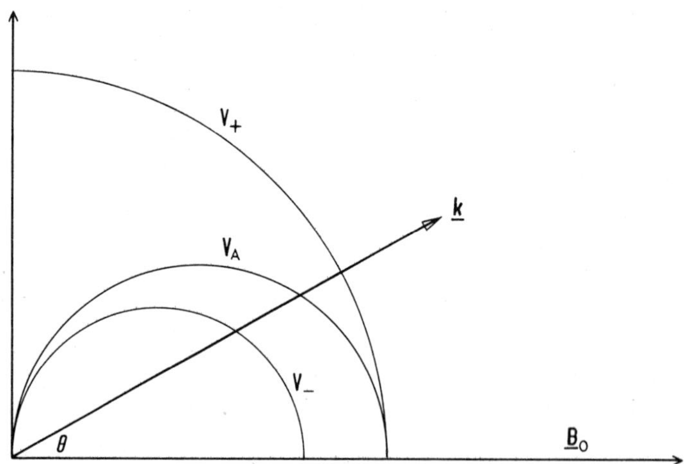

Fig. 3.3 Polar diagram for magnetosonic waves V_+, V_-

Physically, the difference between V_+ and V_- can be seen by comparing the signs of the fluctuation in gas pressure and magnetic energy density, p_1 and $\underline{B}_o \cdot \underline{B}_1/\mu_o$.

$$\frac{\underline{B}_o \cdot \underline{B}_1}{\mu_o} = \left[(\underline{z} \cdot \underline{b})(\underline{v}_1 \cdot \underline{b}) - \underline{z} \cdot \underline{v}_1 \right] \frac{B_o^2}{\mu_o V}$$

$$= \frac{C_A^2}{C_S^2} \left[1 - \frac{C_S^2}{V^2} \cos^2 \theta \right] p_1 , \qquad (3.68)$$

where we have used the results

$$\underline{v}_1 \cdot \underline{b} = - \underline{z} \cdot \underline{b} \, \frac{C_S^2 \, \rho_1}{V \rho_o}$$

$$\underline{z} \cdot \underline{b} = \cos \theta$$

$$\underline{z} \cdot \underline{v}_1 = - \frac{\rho_1 V}{\rho_o}$$

$$p_1 = C_S^2 \, \rho_1 .$$

Hence, p_1 and $\underline{B}_o \cdot \underline{B}_1 / \mu_o$ have the same sign if $V > C_S \cos\theta$, the opposite sign if $V < C_S \cos\theta$. Further, it is straightforward to show that $V_+ > C_S \cos\theta$ and $V_- < C_S \cos\theta$, so the conclusion is that in the fast magnetosonic wave the pressure and magnetic energy fluctuations reinforce each other, the opposite being the case for the slow magnetosonic wave.

3-6 STABILITY

We are concerned in this section with the stability of an equilibrium configuration of an ideally conducting fluid. In §3-4 some examples were given of such equilibria, without discussion of the possibility that small perturbations of the system could grow to such an extent that the equilibrium could not be maintained. Experimentally, violent perturbations have been readily observed in many devices and have led to much interesting theoretical analysis. An outline of some of this work will be given here, though it will be impossible to give a full account. The reader is referred to the bibliography for a more complete description. It should also be mentioned that instabilities may arise for a variety of reasons, including the effect of finite transport coefficients and also due to the fact that a plasma is in reality a collection of interacting ions and electrons. The former are sometimes termed resistive instabilities and can play a very important rôle, though we shall not be able to give them a proper treatment in this book.

The latter form part of the next chapter, which deals with plasma kinetic theory. Due to their highly localised nature, and the fact that they arise from the particle nature of the plasma, they are sometimes known as <u>micro-instabilities</u>.

One method of solving stability problems is that of normal modes, similar to that of dynamical systems with a finite number of degrees of freedom; a fluid has an infinite number of degrees of freedom, but the concept is essentially unaltered. A general perturbation of the plasma does not preserve its form as it evolves in time. Certain perturbations do, and these are the normal modes. In this discussion we consider only <u>small</u> depart-ures from equilibrium, so we shall investigate <u>linear</u> stability. A general perturbation may then be expressed as a linear super-position of normal perturbations. The stability of the system is thus determined by the stability of all the normal modes.

The method of normal modes gives a precise and complete account of the growth rates and oscillation frequencies of normal modes. However, except in the simplest systems, the calculation can be very involved and often yields much more information than that required, which is: Is the equilibrium stable or unstable?

This suggests that a method should be sought which yields this more qualitative information with much less labour. Just as in the dynamics of simple systems we should have recourse to an energy principle, so we can obtain one for perfectly conducting fluids. These then, the theory of normal modes and the energy principle, will be the topics for the next two sections of this chapter. However, before describing these, we shall give an account of an important class of instabilities, known as inter-change instabilities, applicable strictly to systems which close on themselves, as in a torus. A stability criterion can be obtained, based essentially on an energy principle, and although it lacks the mathematical rigour of the later sections, the theory, due to Rosenbluth and Longmire[6], is elegant in its simplicity and wide in application.

3-6-1 Interchange perturbations

The main idea is that we consider two infinitesimal magnetic flux tubes, together with the plasma contained by them. If these tubes are interchanged, then a rearrangement of energy occurs and if the

final energy is lower than that of the initial state, it is
energetically possible for the interchange physically to occur.
The central assumption of the theory is that it will also be
dynamically possible.

Referring to Fig. 3.4, we define, for flux tube $i = 1, 2$,

$$S_i \ = \ \text{cross sectional area}$$

$$B_i \ = \ \text{magnetic field}$$

$$\phi_i \ = \ \text{magnetic flux}$$

$$p_i \ = \ \text{pressure}$$

$$V_i \ = \ \text{volume of flux tube.}$$

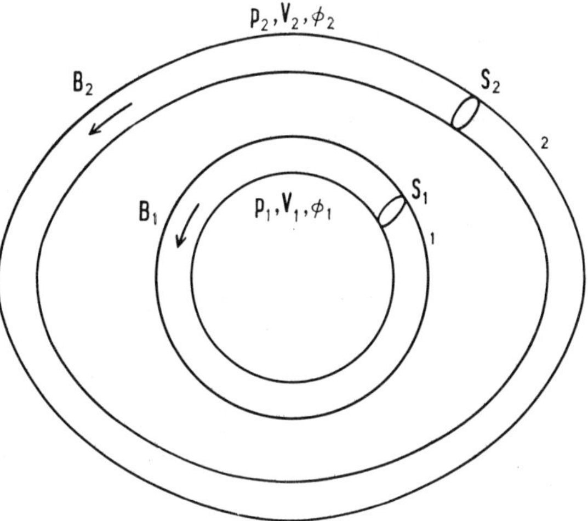

Fig. 3.4 Magnetic flux tubes

$$\phi_i \ = \ B_i S_i$$

$$= \ \text{constant, by definition.}$$

Note that B_i and S_i may vary along the length of the tube.

$$V_i \ = \ \phi_i U_i ,$$

where $U_i = \oint_i \dfrac{dl}{B_i}$, since $V_i = \oint_i S_i \, dl$. $\hspace{3em}$ (3.70)

We now suppose that the flux tubes are interchanged

$$\text{Then } V_1 \rightarrow V_2, \quad V_2 \rightarrow V_1$$

$$B_1 \rightarrow B_2', \quad B_2 \rightarrow B_1'$$

$$p_1 \rightarrow p_2', \quad p_2 \rightarrow p_1'.$$

The final values of the pressures and magnetic fields are defined
by assuming that the plasma expands or compresses adiabatically
and by conserving magnetic flux.   Thus

$$p_1' = p_2 \left(\frac{V_2}{V_1}\right)^\gamma \; ; \qquad p_2' = p_1 \left(\frac{V_1}{V_2}\right)^\gamma$$

$$B_1' \, S_1 = B_2 \, S_2 = \phi_2; \qquad B_2' \, S_2 = B_1 \, S_1 = \phi_1.$$

The total change in energy δW is given by

$$\delta W = \delta W_p + \delta W_B$$

$$\delta W_p = \frac{1}{\gamma-1} \left\{ \left(p_1' V_1 + p_2' V_2\right) - \left(p_1 V_1 + p_2 V_2\right) \right\} \tag{3.71}$$

$$\delta W_B = \frac{1}{2\mu_o} \left\{ \oint_1 B_1'^2 S_1 \, dl + \oint_2 B_2'^2 S_2 \, dl - \oint_1 B_1^2 S_1 \, dl - \oint_2 B_2^2 S_2 \, dl \right\}. \tag{3.72}$$

For adjacent flux tubes, $p_2 = p_1 + \delta p$, $V_2 = V_1 + \delta V$.   We shall
retain only lowest order, i.e. second order in small quantities.
After appropriate expansions and rearranging terms, we have

$$\delta W_p = \delta p \, \delta V + \frac{\gamma p_1}{V_1} (\delta V)^2 \tag{3.73}$$

$$\delta W_B = \tfrac{1}{2}(\phi_2^2 - \phi_1^2) \left[\frac{I_1}{\phi_1} - \frac{I_2}{\phi_2}\right]$$

$$= -\tfrac{1}{2} \, \delta(\phi^2) \, \delta\left(\frac{I}{\phi}\right) , \tag{3.74}$$

where $I = \dfrac{1}{\mu_o} \oint B dl = $ current enclosed by path.   An <u>instability</u>
criterion is then derived by considering the interchange of two
flux tubes containing the same magnetic flux, $\phi_1 = \phi_2$, so that
$\delta\phi = 0$.   As a consequence, $\delta W_B = 0$.   If then $\delta W_p < 0$ energy is
lowered by the interchange and the system is unstable.   The
criterion can be expressed as

$$\delta p \, \delta U + \frac{\gamma p}{U} (\delta U)^2 < 0 \tag{3.75}$$

using $U = \oint \dfrac{dl}{B}$.

(3.75) is clearly only a <u>sufficient</u> condition for instability.
However, there may be other choices of $\delta\phi \neq 0$ which lower δW even
more.   We should more correctly minimise δW with respect to
arbitrary variations in $\delta\phi$ and δp.   However, the result which
follows only slightly modifies the simpler criterion (3.75),
provided

$$\beta = \frac{2\mu_o p}{B^2} \ll 1,$$

which is usually the case in plasma experiments. It is not difficult to understand this result qualitatively. If $\beta \ll 1$, magnetic field effects dominate in δW. Unless large currents flow, these fields are essentially vacuum fields, which are certainly stable, so that any perturbation of these fields necessarily gives $\delta W_B \geqslant 0$, and $\delta W_B = 0$ occurs if and only if $\delta\phi = 0$. These are consequently among the most dangerous perturbations in low β plasmas.

<u>Applications</u>

(i) Plasma in the field of a straight current I_o

The azimuthal magnetic field B is given by

$$B = \frac{\mu_o I_o}{2\pi r} .$$

Hence
$$U = \oint \frac{dl}{B} = \frac{2\pi r}{B} = \frac{(2\pi r)^2}{\mu_o I_o} .$$

We may assume that $p = p(r)$, so that the criterion takes on the simpler form, noting that $\dfrac{dU}{dr} > 0$,

$$\frac{dp}{dr} + \frac{\gamma p}{U}\, \frac{dU}{dr} > 0. \tag{3.76}$$

This gives a critical dependence for the pressure

$$p = p_o\, r^{-2\gamma} .$$

If p decreases more rapidly than this, the plasma will be unstable.

(ii) Plasma in the field of a magnetic dipole

Here, $B \simeq r^{-3}$, while the length of a given field line is proportional to r.   Hence $U \simeq r^4$.   The limiting form for the pressure is now $r^{-4\gamma}$.   We might expect this criterion to be relevant to the Van Allen belts, formed by plasma trapped in the earth's magnetic dipole field.

(iii) Plasma in a toroidal multipole field

In the general multipole system, N internal concentric rings carry current in the same sense.

N = 2 Quadrupole

Fig. 3.5 shows a toroidal cross section of the field with the currents flowing upwards.

$\underline{B}$ is defined in terms of a flux function ψ.   If $\underline{z}$ is a unit vector parallel to the toroidal axis, in the direction of the

45

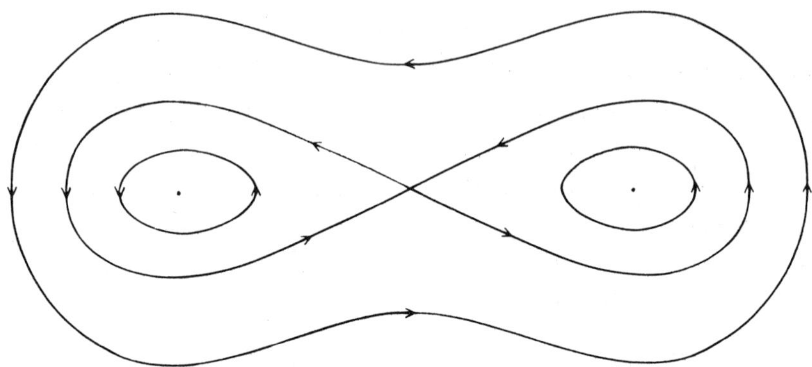

Fig. 3.5   Cross section of toroidal quadrupole field.

current, $\underline{B} = \nabla \times (\psi \, \underline{z})$.

Each field line is characterised by a value of ψ.   The separatrix passes through the null point $B = 0$ and is assigned $\psi = 0$.   The instability criterion can now be written, in terms of ψ, thus

$$\frac{dp}{d\psi} \frac{dU}{d\psi} + \frac{\gamma p}{U} \left(\frac{dU}{d\psi}\right)^2 > 0, \qquad (3.77)$$

where we have used the fact that $p = p(\psi)$ for an equilibrium configuration.   The second term in (3.77) is usually small, and $p(\psi)$ decreases with ψ for containment, so the system is likely to be unstable if $\frac{dU}{d\psi} < 0$.   Calculations have shown that there should be a critical ψ_c at which U has a minimum, $\frac{dU}{d\psi} = 0$, and that <u>inside</u> the corresponding field line the plasma is stable, <u>outside</u> unstable.

This was verified experimentally by Meade[7] (1966), in the case of an octopole field, $N = 4$, which is shown in Fig. 3.6.

In Meade's experiment, hydrogen plasma, at a temperature of 10 eV and density $10^{15} \, \mathrm{m}^{-3}$, was injected into the confining region. Within the stable region bounded by the flux surface ψ_c, electric field fluctuations were very small.   Outside ψ_c fluctuations several orders of magnitude greater were observed.   Density measurements showed a steep density profile just within ψ_c.   Loss of plasma occurred in approximately 600 μs compared with a calculated mean life of 1 ms assuming thermal bombardment of the supports within ψ_c.   Outside ψ_c, the mean lifetime of plasma was around 1 μs, and overall, if there were no stabilising effect due to the magnetic field, the plasma would be expected to have been

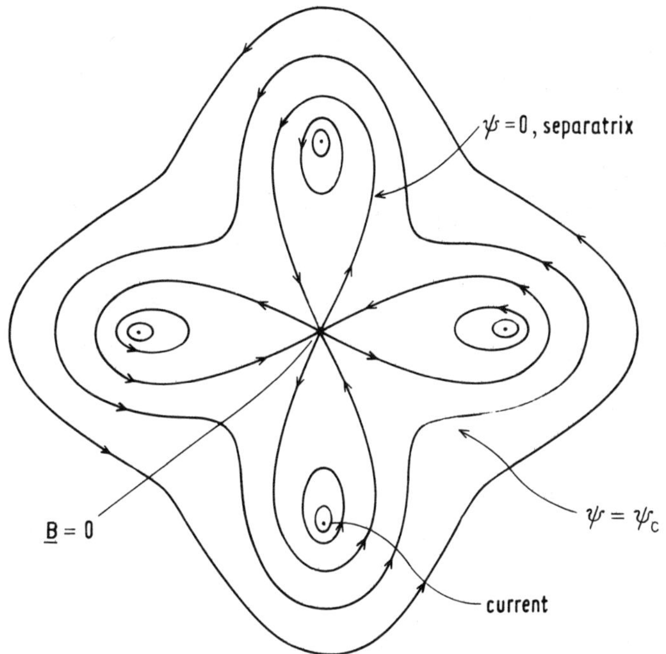

Fig. 3.6 Cross section of toroidal octopole field.
completely lost in about 160 μs.

3-6-2 Method of normal modes

We shall describe one example of normal modes for illustration of
the method, but first discuss briefly boundary conditions which
must be satisfied at an interface of a conducting fluid with a
vacuum, rigid wall, or another conducting fluid.   These boundary
conditions are obtained by integrating the fluid equations and
Maxwell's equations across the boundary.   We shall just quote
results, with comments.

Let $\underline{n}$ be the unit normal to the interface.   Then $\underline{n}.\underline{B}$, $\underline{n}.\underline{v}$ ana
$p + \dfrac{B^2}{2\mu_o}$ are continuous.   These follow, respectively, from the
equation $\nabla.\underline{B} = 0$, the equation of continuity and the fluid equation
of motion, and must be applied at the perturbed interface in the
case of fluid-vacuum and fluid-fluid.   If the boundary is perfectly
conducting $\underline{n}.\underline{B} = 0$, and if it is rigid $\underline{n}.\underline{v} = 0$.   The fluid
equations do not exclude current sheets on the interface, and if
there is a current sheet $\underline{j}*$ we have $\underline{n} \times \left[\underline{B}\right] = \mu_o \underline{j}*$, where $\left[X\right]$
denotes the change in any quantity X in going across the interface.

47

We illustrate some of these points by investigating the stability of a particular model of the cylindrical pinch. This example has been chosen partly because of its mathematical tractability, but also to introduce some of the geometrical complications of cylindrical systems because of their central role in plasma containment.

We therefore investigate the stability of a cylinder of plasma contained within a radius r_o by a cylindrical current sheet in which current flows only axially. There is no magnetic field within the plasma and only an azimuthal field outside, which falls off as $\frac{1}{r}$.

<u>Equilibrium</u>

plasma $p = p_o$ $\rho = \rho_o$
$r < r_o$

$$\text{vacuum} \quad \underline{B}_o = \left(0, \frac{B_o r_o}{r}, 0\right)$$
$r > r_o$

$$\text{boundary} \quad p_o = \frac{B_o^2}{2\mu_o} . \tag{3.78}$$
$r = r_o$

<u>Perburbation</u>

Linearised equations are obtained by setting any variable $q = q_o + q_1$, where q_o is its equilibrium value and q_1 is a small perturbation. The resulting equations, on retaining only lowest order terms, are

$$\text{plasma} \quad \rho_o \frac{\partial \underline{v}_1}{\partial t} = -\nabla p_1 \tag{3.79}$$
$$r < r_o$$

$$\frac{1}{p_o} \frac{\partial p_1}{\partial t} = \frac{\gamma}{\rho_o} \frac{\partial \rho_1}{\partial t} \tag{3.80}$$

$$\frac{\partial \rho_1}{\partial t} + \rho_o \nabla \cdot \underline{v}_1 = 0 \tag{3.81}$$

$$\text{vacuum} \quad \nabla \cdot \underline{B}_1 = 0 \tag{3.82}$$

$$r > r_o \quad \nabla \times \underline{B}_1 = 0 . \tag{3.83}$$

Equation (3.79) results, since $\underline{B}_o$ vanishes for $r < r_o$. The normal modes in θ, z coordinates correspond to a Fourier analysis. In linear theory, on account of the superposition principle, these can be separately analysed. Thus, any first order quantity q_1 may be expressed in the form

$$q_1 \ (r, \ \theta, \ z, \ t) = q_1(r) \ \exp\left[i(m\theta + kz) + \omega t\right]. \tag{3.84}$$

In (3.84), m must be an integer or zero for continuity, $\theta \to \theta + 2\pi$, but k may be any real number.

In particular, we may take the perturbed boundary of the plasma to have the equation

$$r = r_o + r_1 \ \exp\left[i(m\theta + kz) + \omega t\right]. \tag{3.85}$$

Fig. 3.7 shows the distortion (broken line) in the boundary corresponding to m = 0, 1 and 2.

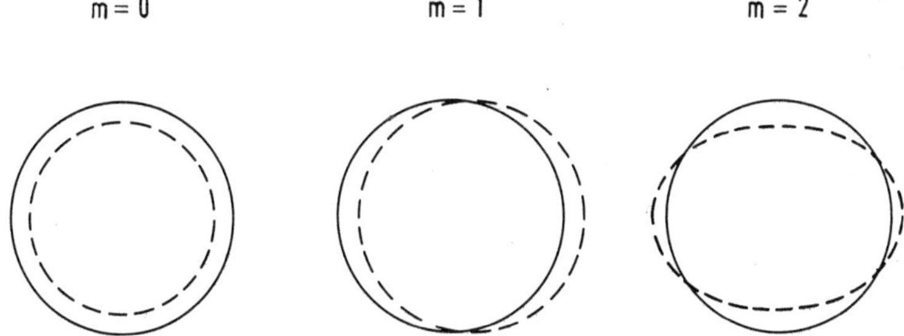

Fig. 3.7   m = 0, 1 and 2 perturbations

We now investigate the stability of each (m, k) mode, by obtaining the general solution of equations (3.79) - (3.83), and applying the boundary conditions.   This results in an equation of the form F (m, k, ω) = 0, the dispersion relation, and an examination of the solution ω of this equation will show if a particular mode is stable (real part of $\omega \leqslant 0$) or unstable (Re(ω) > 0).

$$\underline{\text{Plasma}} \qquad \rho_o \ \omega \underline{v}_1 = -\left(p_1', \ \frac{im}{r} \ p_1, \ ikp_1\right)$$

$$\frac{\omega p_1}{p_o} = \frac{\gamma \omega \rho_1}{\rho_o} \tag{3.86}$$

$$\omega \rho_1 + \rho_o \ \nabla . \underline{v} = 0,$$

where $p_1' = \dfrac{dp_1}{dr}$.

Hence,

$$p_1 = - \ \frac{\gamma p_o}{\omega} \ \nabla . \underline{v}_1$$

$$= \frac{\gamma p_o}{\rho_o \omega^2} \ \left\{\frac{1}{r} \ (rp_1')' - \left[\frac{m^2}{r^2} + k^2\right] \ p_1\right\} . \tag{3.87}$$

Letting $C_s^2 = \gamma p_o / \rho_o$ and $q^2 = k^2 + \omega^2/C_s^2$, equation (3.87) becomes

49

$$\frac{1}{r} (rp_1')' - \left[\frac{m^2}{r^2} + q^2\right] p_1 = 0, \qquad\qquad (3.88)$$

which is a Modified Bessel equation.

Since p_1 is regular at $r = 0$,

$$p_1 = A\, I_m\, (qr), \quad A \text{ arbitrary constant,}$$

and

$$\underline{v}_1 = -\frac{A}{\rho_o \omega} \left[q\, I_m'\, (qr), \; \frac{im}{r}\, I_m\, (qr), \; ik\, I_m\, (qr)\right].$$

<u>Vacuum</u>   Since $\nabla \times \underline{B}_1 = 0$, $\underline{B}_1 = \nabla\psi$, where ψ is the magnetostatic potential, satisfying

$$\nabla^2\psi = 0. \qquad\qquad (3.89)$$

We now want the solution of (3.89), in cylindrical geometry, regular at $r = \infty$, that is

$$\psi = B\, K_m\, (kr), \quad B \text{ arbitrary constant.}$$

Hence,

$$\underline{B}_1 = B \left[kK_m'\, (kr), \; \frac{im}{r}\, K_m\, (kr), \; ik\, K_m\, (kr)\right].$$

<u>Boundary</u>   The normal to a surface defined by $f(r)\,\exp i(m\theta + kz)$ = constant has components proportional to

$$\left(\frac{\partial f}{\partial r}, \; \frac{im}{r}\, f, \; ikf\right).$$

The boundary equation (3.85) thus leads to unit normal

$$\underline{n} = \left(-1, \; im\, \frac{r_1}{r_o}, \; ikr_1\right).$$

Also, the radial component v_{1r} of the velocity of the boundary is $\dfrac{\partial r}{\partial t} = \omega r_1$.   Hence, in (3.85), $r_1 = v_{1r}/\omega$.

We are now in a position to apply the boundary conditions.

(i) $\qquad\qquad\qquad \underline{n}.\underline{B} = 0$

$$B_{1r} = im\, \frac{r_1}{r_o}\, B_{o\theta} = im\, \frac{r_1}{r_o}\, B_o = im\, \frac{v_{1r}}{r_o \omega}\, B_o.$$

Hence,

$$Bk\, K_m'\, (kr_o) = -\frac{imB_o}{r_o \omega^2}\, \frac{A}{\rho_o}\, q\, I_m'\, (qr_o).$$

(ii) $\qquad\qquad\qquad \left[p + \frac{B^2}{2\mu_o}\right] = 0.$

In applying this condition we must take account of the perturbation of the boundary.   Hence, we need to use it in the form

50

$$p_o + p_1 = \frac{B_o^2}{2\mu_o} + \frac{\underline{B}_o \cdot \underline{B}_1}{\mu_o} + \frac{r_1 B_o}{\mu_o} \frac{d}{dr}\left(B_o \frac{r_o}{r}\right)\Bigg|_{r = r_o}$$

$$\text{i.e. } p_1 = \frac{B_o B_{1\theta}}{\mu_o} - \frac{B_o^2}{\mu_o} \frac{r_1}{r_o}$$

$$= \frac{B_o B_{1\theta}}{\mu_o} - \frac{B_o^2}{\mu_o} \frac{v_{1r}}{\omega r_o} .$$

Hence

$$A\, I_m(qr_o) = \frac{B_o}{\mu_o} \frac{im}{r_o} B\, K_m(kr_o) + \frac{B_o^2}{\mu_o \omega^2 r_o^2} \frac{A}{\rho_o} q I_m'(qr_o) .$$

The dispersion relation $F(m, k, \omega) = 0$ is obtained as the eliminant of A and B, namely

$$\begin{vmatrix} \left\{ I_m(qr_o) - \dfrac{B_o^2}{\mu_o \rho_o r_o \omega^2}\, q\, I_m'(qr_o) \right\} - \dfrac{imB_o}{\mu_o r_o} K_m(kr_o) \\[2em] \dfrac{imB_o}{r_o \rho_o \omega^2}\, q\, I_m'(qr_o) \qquad\qquad\qquad k K_m'(kr_o) \end{vmatrix} = 0. \qquad (3.90)$$

(3.90) can be simplified by the introduction of dimensionless parameters (this is always a very useful procedure)

$$x = kr_o$$

$$y = \frac{\omega r_o}{C_s}$$

$$z = qr_o$$

(Note that $z^2 = x^2 + y^2$).

The dispersion relation then becomes

$$\tfrac{1}{2}\gamma y^2 \frac{I_m(z)}{z\, I_m'(z)} = 1 + m^2 \frac{K_m(x)}{x\, K_m'(x)} . \qquad (3.91)$$

We now have to consider all modes (m, k). As y^2 varies from 0 to ∞, LHS > 0 and varies from 0 to ∞. Instability occurs for any (m, k) with a positive y^2. Hence instability occurs for any (m, k) such that RHS > 0, that is

$$1 + \frac{m^2 K_m(x)}{x\, K_m'(x)} > 0. \qquad (3.92)$$

$$m = 0 : \text{ satisfied, hence unstable}$$

$$m = 1 : \text{ unstable for all } x$$

$$m > 1 : \quad \text{for large enough x,} \quad \frac{K_m(x)}{K_m'(x)} \to -1$$

so RHS $\to 1 - \frac{m^2}{x}$ and we now require $1 - \frac{m^2}{x} > 0$. We might therefore expect stability for wave numbers k such that $o \leqslant x \leqslant m^2$.

In fact, for $m > 1$, a more accurate estimate is $0 \leqslant x \leqslant m^2 - 1$, and the directly calculated threshold x_c, using (3.92), is given below for several values of m:

m	2	3	4	5
x_c	3.04	8.02	15.01	24.01

We have deliberately followed this illustration of the technique of normal modes to its conclusion. The precision obtainable is to be noted. However, it should be obvious from this example that it will be a fairly involved process to obtain dispersion relations for any but the simplest equilibrium configurations. Here, the plasma and confining magnetic field are separate. Normally we are interested in interpenetrating plasma and magnetic field. Thus, as we saw in §3-4, the most general configuration possessing cylindrical symmetry is

$$\underline{B}_o = (0, \, B_\theta(r), \, B_z(r)); \quad p_o(r).$$

However, if at the end of the calculation, we are just interested in the regions of stability and instability, it is clear that there is a considerable surplus of information about the actual values of growth rates and oscillation frequencies. It is in this respect that the energy principle method, which we go on to discuss, has proved so powerful a tool in plasma physics research.

3-6-3 The energy principle[8]

In general, a configuration defined in terms of a time-independent magnetic field $\underline{B}_o(\underline{r})$ and a plasma pressure $p_o(\underline{r})$ is in static equilibrium if

$$\nabla p_o = \frac{1}{\mu_o} \, (\nabla \times \underline{B}_o) \times \underline{B}_o. \tag{3.93}$$

We wish to study the behaviour of this plasma configuration when it is slightly perturbed from equilibrium. In the previous section we introduced a perturbation q_1 corresponding to any equilibrium quantity q_o (including variables which vanish in equilibrium, such as the velocity and the electric field). We first show that all perturbed quantities can be determined in terms of a displacement vector field $\underline{\xi}\,(\underline{r},\,t)$.   Thus,

$$\underline{v}_1 \;=\; \frac{\partial \underline{\xi}}{\partial t}\,. \tag{3.94}$$

Then, since $\nabla \times \underline{E}_1 = -\dfrac{\partial \underline{B}_1}{\partial t}$ and $\underline{E}_1 + \underline{v}_1 \times \underline{B}_o = 0$,

$$\frac{\partial \underline{B}_1}{\partial t} \;=\; \nabla \times (\underline{v}_1 \times \underline{B}_o),$$

so that, using (3.94), we obtain

$$\underline{B}_1 \;=\; \nabla \times (\underline{\xi} \times \underline{B}_o). \tag{3.95}$$

Similarly, $\dfrac{1}{p_o}\,\dfrac{\partial p_1}{\partial t} = \dfrac{\gamma}{\rho_o}\,\dfrac{\partial \rho_1}{\partial t} = -\dfrac{\gamma}{\rho_o}\,\nabla \cdot (\rho_o\,\underline{v}_1).$

Hence

$$\frac{p_1}{p_o} \;=\; -\frac{\gamma}{\rho_o}\,\nabla \cdot (\rho_o\,\underline{\xi})$$

and using $p_o \rho_o^{-\gamma} = $ constant, we finally obtain

$$p_1 \;=\; -\gamma p_o \nabla\cdot\underline{\xi} - \underline{\xi}\cdot\nabla p_o. \tag{3.96}$$

Using (3.95) and (3.96), we can express the fluid equation of motion as an equation for $\underline{\xi}$, thus:

$$\rho_o \frac{\partial \underline{v}_1}{\partial t} \;=\; -\nabla p_1 + \frac{1}{\mu_o}\left[(\nabla \times \underline{B}_1) \times \underline{B}_o + (\nabla \times \underline{B}_o) \times \underline{B}_1\right]..$$

Hence

$$\rho_o \frac{\partial^2 \underline{\xi}}{\partial t^2} \;=\; \nabla\left[\gamma p_o \nabla\cdot\underline{\xi} + \underline{\xi}\cdot\nabla p_o\right] +$$

$$+ \frac{1}{\mu_o}\left[\nabla \times \nabla \times (\underline{\xi} \times \underline{B}_o)\right] \times \underline{B}_o +$$

$$+ \frac{1}{\mu_o}(\nabla \times \underline{B}_o) \times \left[\nabla \times (\underline{\xi} \times \underline{B}_o)\right]. \tag{3.97}$$

The next step is to introduce a Fourier analysis of the time-dependence in (3.97).   Substitution of

$$\underline{\xi}\,(\underline{r},\,t) \;=\; \int \underline{\xi}\,(\underline{r},\,\omega)e^{i\omega t}\,d\omega$$

into (3.97) leads to an equation for $\underline{\xi}\,(\underline{r},\,\omega)$ of the form

$$-\rho_o \omega^2\,\underline{\xi} \;=\; F\,\underline{\xi}, \tag{3.98}$$

where F is a linear operator acting on $\underline{\xi}$.   Equation (3.98) is in fact the normal mode equation for $\underline{\xi}\,(\underline{r})$, which, when combined with appropriate boundary conditions on $\underline{\xi}$, forms an eigenvalue problem for ω.

<u>Boundary conditions</u>

(i) $\underline{\xi}$ everywhere finite

(ii) $\underline{n}.\underline{\xi} = 0$ at a rigid wall,

    where $\underline{n}$ is the unit normal at the boundary.

These boundary conditions are appropriate to the case of a plasma extending right out to the boundary, so that there is no vacuum region external to the plasma, as in the problem discussed in the previous section.   Comparison theorems can be proved which allow comparison between the stability of a plasma configuration bounded by a rigid wall and one with an external vacuum region. The result is that the plasma surrounded by a vacuum region will always be as stable as one surrounded by a low density ideal plasma.

In the case of a cylindrical plasma, in which the equilibrium state possesses cylindrical symmetry, $\underline{\xi}$ may be further analysed as

$$\underline{\xi}\,(\underline{r}) \;=\; \underline{\xi}\,(r)\,\exp\,i(m\theta + kz). \tag{3.99}$$

The equation of motion reduces to an equation for $\underline{\xi}\,(r)$, and application of boundary conditions eventually leads to a dispersion relation

$$F\,(m,\,k,\,\omega) \;=\; 0$$

as before.   Results will be identical with any other normal mode calculation, if we follow this procedure, except that the treatment is systematic, with $\underline{\xi}(r)$ always the fundamental perturbation.

However, the whole point of introducing $\underline{\xi}$ is that it allows the introduction of a variational principle.

We first form the scalar product of $\underline{\xi}$ with equation (3.98), and integrate over the entire plasma volume.

$$-\omega^2 \int \rho_o\;\underline{\xi}.\underline{\xi}\;\;d\underline{r}\;=\;\int \underline{\xi}\;.\;F\,\underline{\xi}\;d\underline{r}.$$

$$\text{Then,}\quad \omega^2 \;=\; \frac{2\,\delta W\,(\underline{\xi},\,\underline{\xi})}{\int \rho_o\underline{\xi}.\underline{\xi}\;d\underline{r}}\;, \tag{3.100}$$

where $\quad \delta W \;=\; -\tfrac{1}{2}\int \underline{\xi}.F\underline{\xi}\;d\underline{r}$

$$= \; \tfrac{1}{2}\int \left[\frac{1}{\mu_o}\,\underline{Q}.\underline{Q} - \frac{1}{\mu_o}\,\nabla \times \underline{B}_o.\underline{Q} \times \underline{\xi} + \gamma p_o\,(\nabla.\underline{\xi})^2 + \nabla.\underline{\xi}\,(\underline{\xi}.\nabla p_o) \right]\,d\underline{r}$$

$$\tag{3.101}$$

and $\quad \underline{Q} \;=\; \nabla \times (\underline{\xi} \times \underline{B}_o) \;=\; \underline{B}_1 \quad (\text{cf. } (3.95)).$

This final form (3.101) for δW arises from integration by parts and applying the boundary condition at the wall, $\underline{n}.\underline{\xi} = 0$. For example,

$$-\int \underline{\xi}.(\nabla \times \underline{Q}) \times \underline{B}_o \, d\underline{r} \;=\; \int (\nabla \times \underline{Q}) \,.\, \underline{\xi} \times \underline{B}_o \, d\underline{r}$$

$$= \int \epsilon_{ijk} \frac{\partial Q_j}{\partial x_i} (\underline{\xi} \times \underline{B}_o)_k \, d\underline{r}$$

$$= -\int \epsilon_{ijk} Q_j \frac{\partial}{\partial x_i} (\underline{\xi} \times \underline{B}_o)_k \, d\underline{r}$$

$$= \int Q_j \left[\nabla \times (\underline{\xi} \times \underline{B}_o) \right]_j \, d\underline{r}$$

$$= \int \underline{Q}.\underline{Q} \, d\underline{r}.$$

Note that, in δW, $\dot{\underline{\xi}}$ does not appear, only $\underline{\xi}$.

We now apply the Rayleigh principle, that obtaining a turning value of the RHS of equation (3.100) with respect to arbitrary variations in $\underline{\xi}$, subject only to the boundary conditions, is entirely equivalent to solving the normal mode equation (3.98). In other words, a solution of (3.98) when inserted in the RHS of (3.100) will give a turning value, and vice versa.   We write this principle as

$$\Delta_{\underline{\xi}} \left(\frac{2\delta W}{\int \rho_o \, \underline{\xi}.\underline{\xi} \, d\underline{r}} \right) \;=\; 0. \tag{3.102}$$

To prove the principle (3.102), we require F to be a self-adjoint linear operator.   That is, we first of all write $F\underline{\xi}$ in component form, namely $(F\underline{\xi})_i = F_{ij}\,\xi_j$, summation over j implied. Then it can be shown that, for any arbitrary vector $\underline{\eta}$ satisfying the same boundary conditions as $\underline{\xi}$

$$\int d\underline{r}\, \xi_i\, F_{ij}\, \eta_j \;=\; \int d\underline{r}\, \eta_i\, F_{ij}\, \xi_j$$

$$\text{or} \quad \int d\underline{r}\, \underline{\xi} \,.\, F\underline{\eta} \;=\; \int d\underline{r}\, \underline{\eta} \,.\, F\,\underline{\xi}.$$

We now suppose $\underline{\xi} \to \underline{\xi} + \delta\underline{\xi}$ in (3.102). Then,

$$\frac{\int (\delta\underline{\xi}.F\underline{\xi} + \underline{\xi}.F\delta\underline{\xi})d\underline{r}}{\int \rho_o \, \underline{\xi}.\underline{\xi} \, d\underline{r}} \;=\; \frac{\int \underline{\xi}.F\underline{\xi} \, d\underline{r} \,.\, 2\int \rho_o \, \delta\underline{\xi}.\underline{\xi} \, d\underline{r}}{\left(\int \rho_o \, \underline{\xi}.\underline{\xi} \, d\underline{r}\right)^2}.$$

Since F is self-adjoint, the LHS can be written as

$$\frac{2 \int \delta\underline{\xi} \cdot F\underline{\xi} \; d\underline{r}}{\int \rho_0 \; \underline{\xi} \cdot \underline{\xi} \; d\underline{r}}$$

and the denominator is positive definite, so we require

$$\int \delta\underline{\xi} \cdot F\underline{\xi} \; d\underline{r} = \left\{ \frac{\int \underline{\xi} \cdot F\underline{\xi} \; d\underline{r}}{\int \rho_0 \underline{\xi} \cdot \underline{\xi} \; d\underline{r}} \right\} \int \rho_0 \; \delta\underline{\xi} \cdot \underline{\xi} \; d\underline{r} \qquad (3.103)$$

for arbitrary $\delta\underline{\xi}$.

Defining $-\lambda$ to be the term on RHS of (3.103) enclosed by curly brackets, we have

$$\int \delta\underline{\xi} \cdot (\lambda\rho_0 \; \underline{\xi} + F\underline{\xi}) \;\; = \;\; 0,$$

and, since $\delta\underline{\xi}$ is arbitrary, this requires

$$-\lambda\rho_0 \; \underline{\xi} \;\; = \;\; F\underline{\xi},$$

which is the eigenvalue equation (3.98) with $\lambda = \omega^2$ for a non-trivial solution.

We note that the Rayleigh principle gives as much information as the normal mode approach. However, we can adopt a simpler "energy principle" by demanding only a knowledge of the <u>sign</u> of ω^2, and not its absolute magnitude. This is achieved by minimizing δW, and relaxing the normalizing effect of the denominator, $\int \rho_0 \; \underline{\xi} \cdot \underline{\xi} \; d\underline{r}$.

$\Delta(\delta W) = 0$ will not lead to the equation of motion (3.98). We shall show that, if δW (minimum) > 0, then all eigenvalues $\omega^2 > 0$, that is, all modes are purely oscillatory. On the other hand, if δW (minimum) < 0 at least one mode has $\omega^2 < 0$, which implies that that particular mode will in general suffer exponential growth, our criterion for linear instability.

We first discuss the physical meaning of δW, by forming the scalar product of $\dfrac{\partial \underline{\xi}}{\partial t}$ and the equation of motion (3.98), then integrating over $\underline{r}$, t.

$$\int d\underline{r} \; dt \; \rho_0 \; \frac{\partial \underline{\xi}}{\partial t} \cdot \frac{\partial^2 \underline{\xi}}{\partial t^2} = \int d\underline{r} \; dt \; \frac{\partial \underline{\xi}}{\partial t} \cdot F\underline{\xi}$$

$$= \tfrac{1}{2} \int dt \; d\underline{r} \left[\frac{\partial \underline{\xi}}{\partial t} \cdot F\underline{\xi} + \underline{\xi} \cdot F \frac{\partial \underline{\xi}}{\partial t} \right]$$

$$= \tfrac{1}{2} \int dt \; \frac{\partial}{\partial t} \int d\underline{r} \; \underline{\xi} \cdot F\underline{\xi}$$

$$= \tfrac{1}{2} \int d\underline{r} \; \underline{\xi} \cdot F\underline{\xi}$$

$$= -\delta W. \qquad (3.104)$$

LHS of (3.104) is just

$$\int \tfrac{1}{2}\rho_o \left(\frac{\partial \xi}{\partial t}\right)^2 \, d\underline{r} = \text{kinetic energy of perturbation, K.E.}$$

Hence

$$\text{K.E.} + \delta W = 0,$$

and we may interpret δW as the potential energy of the perturbation. Then $\delta W \gtrless 0$ is the same condition as one obtains in the discussion of the stability of dynamical systems with a finite number of degrees of freedom.

Suppose now we have a normal mode solution

$$\underline{\xi}_n \,(\underline{r},\ t) = \underline{\xi}_n \,(\underline{r}) \, \exp(i\omega_n t),$$

which satisfies the eigenvalue equation

$$-\omega_n^2 \, \rho_o \, \underline{\xi}_n \,(\underline{r}) = F \, \underline{\xi}_n \,(\underline{r}). \tag{3.105}$$

Because F is self-adjoint, $\underline{\xi}_n$ can be chosen to satisfy the orthonormal criterion

$$\int d\underline{r} \, \rho_o \, \underline{\xi}_n \cdot \underline{\xi}_{n'} = \delta_{nn'}$$

since

$$-\int \underline{\xi}_{n'} \, \omega_n^2 \, \rho_o \, \underline{\xi}_n \, d\underline{r} = \int \underline{\xi}_{n'} \cdot F \, \underline{\xi}_n \, d\underline{r}$$

$$= \int \underline{\xi}_n \cdot F \, \underline{\xi}_{n'} \, d\underline{r}$$

$$= -\omega_{n'}^2 \int \rho_o \, \underline{\xi}_n \cdot \underline{\xi}_{n'} \, d\underline{r}$$

(provided $\omega_n \neq \omega_{n'}$ if $n \neq n'$).

Reality of eigenvalues can easily be shown. F and ρ_o are real quantities, but for the moment let us assume ω_n^2 is complex, as well as $\underline{\xi}_n$. Then,

$$-\omega_n^2 \, \rho_o \, \underline{\xi}_n = F \, \underline{\xi}_n$$

$$-\omega_n^{2*} \, \rho_o \, \underline{\xi}_n^* = F \, \underline{\xi}_n^*$$

Hence, $\quad -\omega_n^2 \int \rho_o \, \underline{\xi}_n^* \cdot \underline{\xi}_n \, d\underline{r} = \int \underline{\xi}_n^* \cdot F \, \underline{\xi}_n \, d\underline{r}$

$$= \int \underline{\xi}_n \cdot F \, \underline{\xi}_n^* \, d\underline{r}$$

$$= -\omega_n^{2*} \int \rho_o \, \underline{\xi}_n \cdot \underline{\xi}_n^* \, d\underline{r},$$

which shows that $\omega_n^{2*} = \omega_n^2$.

We need only assume that $\underline{\xi}_n$ form a complete set, to obtain the result that the sign of δW_{min} determines the stability of the system.   Clearly, instability of a particular mode n is assured if $\omega_n^2 < 0$, and the completeness property assumed shows that the necessary and sufficient condition for instability is the existence of at least one negative ω_n^2.

Physically, if $\delta W_{min} < 0$, we expect the system to be unstable and at least one negative ω_n^2.   Let $\underline{\xi}$ be a displacement for which $\delta W < 0$ and which satisfies the boundary conditions.   Then,

$$\underline{\xi} \;=\; \sum_n a_n \, \underline{\xi}_n, \qquad \text{all } a_n, \; \underline{\xi}_n \text{ real}$$

$$\delta W \;=\; -\tfrac{1}{2} \sum_{n,n'} a_n \, a_{n'} \int d\underline{r} \; \underline{\xi}_n \cdot F \, \underline{\xi}_{n'}$$

$$=\; \tfrac{1}{2} \sum a_n^2 \, \omega_n^2.$$

This demonstrates clearly that $\delta W < 0$ if and only if at least one $\omega_n^2 < 0$.   Hence examination of the sign of δW for a given possible $\underline{\xi}$ will determine whether the displacement is stable or unstable.   Following this procedure for arbitrary $\underline{\xi}$ will then give the required instability criterion $\delta W_{min} < 0$.

Note that our choice of normalisation

$$\int \rho_o \, \underline{\xi}_n \cdot \underline{\xi}_n \, d\underline{r} \;=\; 1$$

in the context of the Rayleigh principle shows that we gain in simplicity but lose all knowledge of the actual magnitude of ω_n^2.

3-6-4   The Suydam criterion[9]

The energy principle derived in §3-6-3 is applicable to any general static plasma equilibrium, independent of any particular choice of geometry.   In this section, the energy principle is applied to equilibria possessing cylindrical symmetry, and extending right out to a rigid wall at $r = R$.

<u>Equilibrium</u> $0 \leqslant r \leqslant R$

$$\underline{B}_o \;=\; (0, \; B_\theta(r), \; B_z(r)) \tag{3.106}$$

$$p_o \;=\; p_o(r)$$

$$\mu_o p_o' + B_\theta B_\theta' + \frac{B_\theta^2}{r} + B_z B_z' \;=\; 0.$$

where $'$ signifies $\dfrac{d}{dr}$.

<u>Perturbation</u>

$$\underline{\xi}\,(\underline{r}) \;=\; \underline{\xi}\,(r)\,\exp\,i\,(m\theta + kz).$$

We proceed to evaluate, term by term, the integrand of δW, (cf. expression (3.101)), noting that

$$\underline{Q} \;=\; \nabla \times (\underline{\xi} \times \underline{B}_o)$$

$$= \left[\,i\alpha\xi_r,\; i\alpha\xi_\theta - B_\theta \nabla\cdot\underline{\xi} - r\xi_r\left(\frac{B_\theta}{r}\right)',\; i\alpha\xi_z - B_z\nabla\cdot\underline{\xi} - \xi_r B_z'\,\right], (3.107)$$

where $\alpha = \dfrac{mB_\theta}{r} + kB_z$.

Since we are using an exponential Fourier decomposition for $\underline{\xi}$, we must preserve the reality of δW, so that $\underline{Q}^2 \to \underline{Q}^* \cdot \underline{Q}, \dots$ Inserting (3.107) for $\underline{Q}$ into δW, we can easily carry out the θ, z integrations, yielding, apart from a constant scale factor.

$$\delta W = \int_0^R rdr\left\{\alpha^2\underline{\xi}\cdot\underline{\xi}^* + \left(B_\theta^2 + B_z^2 + \gamma\mu_o p_o\right)\nabla\cdot\underline{\xi}\,\nabla\cdot\underline{\xi}^* -\right.$$

$$-2B_\theta\left(\frac{B_\theta}{r}\right)'\,\xi_r\xi_r^* - \frac{2B_\theta^2}{r}\,(\xi_r\nabla\cdot\underline{\xi}^* + \xi_r^*\nabla\cdot\underline{\xi}) +$$

$$+ i\alpha\left[\frac{2B_\theta}{r}\,(\xi_r^*\,\xi_\theta - \xi_\theta^*\,\xi_r) - B_\theta\,(\xi_\theta\nabla\cdot\underline{\xi}^* - \xi^*\nabla\cdot\underline{\xi}) -\right.$$

$$\left.\left.- B_z\,(\xi_z\nabla\cdot\underline{\xi}^* - \xi_z^*\nabla\cdot\underline{\xi})\right]\right\}\;. \tag{3.108}$$

In obtaining (3.108), we have eliminated p_o', using the equilibrium condition (3.106).

δW is a functional of ξ_r', ξ_r, ξ_θ, ξ_z, but not of ξ_θ', ξ_z'. Hence the Euler-Lagrange equations, obtained by minimizing δW with respect to ξ_θ, ξ_z variations, are purely algebraic. (In general, the Euler equations will determine only an extremum of δW, but in this case we can show by considering higher order terms that δW is in fact minimized.)

Writing δW in the form

$$\delta W \;=\; \int_0^R F\,(\xi_\theta\,\xi_\theta^*\,\xi_z\,\xi_z^*\,\dots\,)$$

we obtain the Euler equations in the form

$$\frac{\partial F}{\partial \xi_\theta} \;=\; 0; \qquad \frac{\partial F}{\partial \xi_z} \;=\; 0.$$

Noting that $\nabla\cdot\underline{\xi} = \dfrac{1}{r}\,(r\xi_r)' + \dfrac{im}{r}\,\xi_\theta + ik\xi_z$ we obtain

$$\frac{\partial F}{\partial \xi_\theta} = \alpha^2 \xi_\theta^* + (B_\theta^2 + B_z^2 + \gamma\mu_o p_o) \frac{im}{r} \nabla . \underline{\xi}^* - \frac{2im}{r^2} B_\theta^2 \xi_r^* +$$

$$+ i\alpha \left[\frac{2B_\theta}{r} \xi_r^* - B_\theta \nabla . \underline{\xi}^* + \frac{im}{r} B_\theta \xi_\theta^* + \frac{im}{r} B_z \xi_z^* \right] = 0.$$

$$(3.109)$$

$$\frac{\partial F}{\partial \xi_z} = \alpha^2 \xi_z^* + ik(B_\theta^2 + B_z^2 + \gamma\mu_o p_o) \nabla . \underline{\xi}^* - \frac{2ik}{r} B_\theta^2 \xi_r^* +$$

$$+ i\alpha \left[-B_z \nabla . \underline{\xi}^* + ik B_\theta \xi_\theta^* + ik B_z \xi_z^* \right] = 0. \qquad (3.110)$$

An immediate consequence of (3.109) and (3.110) can be obtained by combining $B_\theta \times$ (3.109) and $B_z \times$ (3.110), giving

$$p_o \, \alpha \, \nabla . \underline{\xi}^* = 0. \qquad (3.111)$$

Hence, the ξ_θ, ξ_z which minimize δW also, in general, make $\nabla . \underline{\xi}$ vanish, that is, they are incompressible perturbations. The exceptions are cases where $p_o = 0$ or $\alpha = 0$. The first possibility cannot occur, since p_o is positive throughout. However, $\alpha = 0$, or $mB_\theta/r + kB_z = 0$, is possible, and this plays an important rôle, as we shall see later. Usually, though, α will vanish at most at a discrete set of values of r, remembering that m and k are fixed. Hence, by a continuity argument, we shall expect $\nabla . \underline{\xi}$ to vanish everywhere in the range (0, R). A special case occurs if B_θ/rB_z is constant over a finite region, so that the magnetic field, which is helical for any given r, has constant pitch over a finite region.

We now return to equations (3.109) and (3.110), which can be considerably simplified by setting $\nabla . \underline{\xi} = \nabla . \underline{\xi}^* = 0$. It is now a trivial matter to eliminate ξ_θ, ξ_z from δW, in expression (3.108). Define $\xi = \xi_r$ = real variable, and $\beta = kB_z - mB_\theta/r$.

$$\delta W = \int_0^R rdr \left\{ \xi^2 \left[\alpha^2 - \frac{2B_\theta}{r^2} (rB_\theta)' + \frac{\beta^2}{m^2 + k^2 r^2} \right] + \right.$$

$$\left. + 2r\xi\xi' \frac{\alpha\beta}{m^2 + k^2 r^2} + r^2 \xi'^2 \frac{\alpha^2}{m^2 + k^2 r^2} \right\}$$

$$= \int_0^R rdr \left\{ \xi^2 \left[\alpha^2 - \frac{2B_\theta}{r^2} (rB_\theta)' \right] + \frac{1}{m^2 + k^2 r^2} \left[\beta\xi + \alpha r\xi' \right]^2 \right\} .$$

$$(3.112)$$

Both these expressions for δW are useful in the discussion which follows.

Note that $\delta W \gtrless 0$ unless $B_\theta (r B_\theta)' > 0$. Hence complete stability is predicted if B_θ falls off everywhere in the plasma at least as rapidly as $\frac{1}{r}$. We should therefore expect stability if B_θ is due to a current along the axis. This is essentially the basis for certain plasma experimental devices termed 'hardcore' and 'Levitron', the latter being the toroidal counterpart of hardcore cylindrical devices, and possessing a magnetically levitated central ring.

The condition above on B_θ is obviously a sufficient condition.

We can deduce some useful results which allow a simple comparison between the stability characteristics of different (m, k) modes.

Case 1.   m = 0

It is straightforward to prove that, if m = 0,

$$\delta W = \delta W_o + k^2 \int_0^R rdr\, B_z^2\, \xi^2,$$

$$\text{where}\quad \delta W_o = \int_0^R dr \left[r B_z^2\, \xi'^2 + \left(\frac{B_z^2}{r} + 2\mu_o p_o' \right) \xi^2 \right].$$

Hence, k^2 appears only with a positive definite term, and so, for m = 0, stability for $k \to 0$ implies stability for all $|k| > 0$. We must exclude the special case m = 0, k = 0, which is the case of radial oscillations of the plasma.

Case 2.   m > 0

Consider displacements with different m, k such that q = k/m = constant. For such displacements, m occurs only in the term $\frac{m^2}{r} (qr\, B_z + B_\theta)^2$. Since this term is positive definite, then of all modes (m, k) with the same value of q, the least stable is $|m| = 1$. So, if the system is stable for $|m| = 1$, $-\infty < k < \infty$, it is also stable for all $|m| > 1$.

In summary, if the system is stable for m = 0, $k \to 0$, and $|m| = 1$, $-\infty < k < \infty$, it is stable for all (m, k) (cf. results obtained in the simple problem discussed in §3-6-2). We now describe a method of obtaining a necessary condition for stability, known as the Suydam criterion. Let us write the energy δW in the form

$$\delta W = \int_0^R F (r, \xi, \xi')\, dr.$$

As ξ is varied, an extremum, not necessarily a minimum, in δW

is obtained if ξ satisfies Euler's equation

$$\frac{\partial F}{\partial \xi} = \frac{d}{dr}\left(\frac{\partial F}{\partial \xi'}\right),$$

which is of the form

$$\xi'' + P\xi' + Q\xi = 0, \qquad\qquad (3.113)$$

where

$$P = \frac{3}{r} + \frac{2\alpha'}{\alpha} - \frac{2k^2 r}{m^2 + k^2 r^2}$$

$$Q = -\left(k^2 + \frac{m^2 - 1}{r^2}\right) - \frac{2k^2\beta}{\alpha(m^2+k^2 r^2)} - \frac{2\mu_0 k^2 p_0'}{r\alpha^2}.$$

When ξ is a solution of Euler's equation, the integrand of δW is a perfect differential. For

$$\int_0^R F dr = \int_0^R (A\xi^2 + 2H\xi\xi' + B\xi'^2)\,dr$$

and $\quad A\xi + H\xi' = (H\xi + B\xi')'.$

Thus,
$$\int_{r_1}^{r_2} F dr = \int_{r_1}^{r_2} dr\left[\xi\,(A\xi + H\xi') + \xi'\,(H\xi + B\xi')\right]$$

$$= \int_{r_1}^{r_2} dr\left[\xi\,(H\xi + B\xi')' + \xi'\,(H\xi + B\xi')\right]$$

$$= \left[\xi\,(H\xi + B\xi')\right]_{r_1}^{r_2}.$$

The boundary conditions on $\xi(r)$ must now be stated explicitly:

$$\xi(0) = 0, \qquad |m| \neq 1$$
$$\xi(0) \text{ finite}, \qquad |m| = 1$$
$$\xi(R) = 0, \qquad \text{all } m.$$

Further, since ξ describes a physically realisable radial displacement of the plasma, ξ must be finite everywhere in the range $(0, R)$. We would then deduce a straightforward (and trivial) result if ξ, defined by the Euler equation and the boundary conditions, remained finite everywhere. For it is obvious that, except perhaps for $m = 1$, δW would be identically zero. This must be incorrect. The reason is that P and Q, in (3.113) are singular where $\alpha = 0$, so that a solution of the Euler equation will also be singular at such points, which always exist for any non-zero m, since k is in the range $(-\infty, \infty)$. $m = 0$ requires special treatment, since then $\alpha = 0$ implies $k = 0$, and as we have already indicated $m = 0$, $k = 0$ corresponds to a radial pulsation. We shall consider here only $m > 0$.

Suppose that $\alpha = 0$ at $r = a$. In the neighbourhood of a, let the solution ξ_E of Euler's equation (3.113) have a leading term $\xi_o (r - a)^\nu$, where ν is some complex number with at least a negative real part. The indicial equation for ν is

$$\nu(\nu - 1) + 2\nu - \left.\frac{2\mu_o k^2 p'_o}{a\alpha'^2}\right|_{r\,=\,a} = 0.$$

Now let $\mu = B_\theta / r B_z$, so that $\alpha = B_z(m\mu + k)$ and $m\mu + k = 0$ at $r = a$.

$$\text{Thus,}\quad \left.\alpha'\right|_{r\,=\,a} = \left.m\mu'B_z\right|_{r\,=\,a} = \left.\frac{-k\mu'}{\mu}B_z\right|_{r\,=\,a},$$

and the indicial equation is

$$\nu^2 + \nu + M^2 = 0,$$

$$\text{where}\quad M^2 = -\left.\frac{2\mu_o p'_o}{rB_z^2}\left(\frac{\mu}{\mu'}\right)^2\right|_{r\,=\,a}.$$

Hence, the roots of the indicial equation are

$$\nu = -\tfrac{1}{2} \pm \tfrac{1}{2}(1 - 4M^2)^{\frac{1}{2}}. \tag{3.114}$$

If $4M^2 > 1$, the roots are complex and both have negative real part.

If $4M^2 \leqslant 1$, the roots are real and one at least is negative.

Thus, ξ_E is in general singular at $r = a$, since it will be a linear combination of both solutions. It is therefore in general non-physical.

We now construct a physical ξ in the following manner. Given a small length $\varepsilon \ll a$,

$$0 \leqslant r \leqslant a-\varepsilon \qquad \xi = \xi_{E1}$$

$$a-\varepsilon \leqslant r \leqslant a+\varepsilon \qquad \xi = \xi_o = \text{constant}$$

$$a+\varepsilon \leqslant r \leqslant R \qquad \xi = \xi_{E2}.$$

In this definition of a possible physical ξ, ξ_{E1} denotes an Euler solution which satisfies the boundary condition at $r = 0$, while ξ_{E2} satisfies the condition at $r = R$. Further, $\xi_{E1} = \xi_o$ at $r = a -\varepsilon$ and $\xi_{E2} = \xi_o$ at $r = a +\varepsilon$. Two examples of a possible ξ are given in Fig. 3.8.

Note

(i) ξ is continuous and finite everywhere.

(ii) Regions $(0, a)$ and (a, R) are essentially disconnected, so that we no longer have an eigenvalue problem.

We can now evaluate δW in terms of ε and ξ_o.

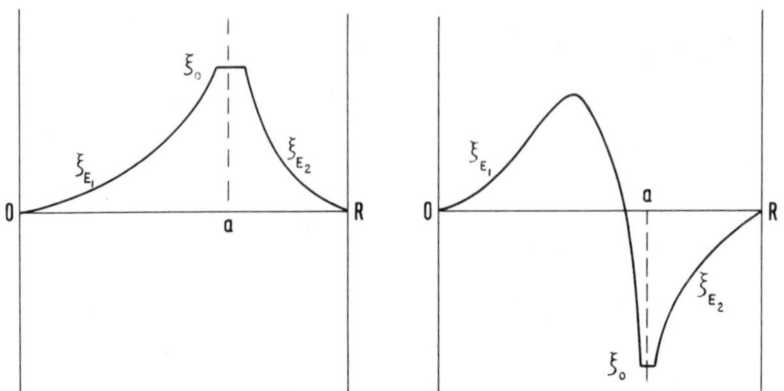

Fig. 3.8 Possible perturbations which are close to Euler solutions.

Case 1. $4M^2 \leqslant 1$ ν real

$$\delta W = 2a\varepsilon\xi_o^2 \left[\frac{|\nu|}{m^2 + k^2 a^2} + \frac{2\mu_o \mu^2 p_o'}{1 + \mu^2 a^2} \right]_{r = a}.$$

This is obtained by evaluating δW in each region

$(0, a - \varepsilon)$, $(a - \varepsilon, a + \varepsilon)$, $(a + \varepsilon, R)$,

$$\delta W = \delta W_1 + \delta W_2 + \delta W_3$$

$$\delta W_1 = \int_0^{a-\varepsilon} F(r, \xi_{E1}) dr = \left[\xi_{E1} (H\xi_{E1} + B\xi_{E1}') \right]_{r = a - \varepsilon}$$

$$\delta W_2 \simeq 2\varepsilon F(a, \xi_o) = 2\varepsilon A\xi_o^2 \Big|_{r = a}$$

$$\delta W_3 = - \left[\xi_{E2} (H\xi_{E2} + B\xi_{E2}') \right]_{r = a + \varepsilon}.$$

Since we know $\xi_E \simeq \xi_o (r - a)^\nu$ near $r = a$, we can get ξ_E and ξ_E' at $r = a \pm \varepsilon$.   This leads immediately to the given expressions for δW_1 and δW_3.   Finally,

$$\delta W = 2a\varepsilon\xi_o \frac{a^2 B_z^2 \mu'^2}{1 + \mu^2 a^2} (|\nu| - M^2),$$

which is always positive, since $|\nu| > M^2$ in this case.   (To show this, note that $|\nu| = \frac{1}{2} \pm \frac{1}{2} (1 - 4M^2)^{\frac{1}{2}}$, hence $|\nu| - M^2 = \frac{1}{4}(1 \pm (1 - 4M^2)^{\frac{1}{2}})^2 \geqslant 0$.)

Hence Case 1 is always stable.

Case 2. $4M^2 > 1$ ν complex

Let $\nu_{1,2} = -\frac{1}{2} \pm \frac{1}{2}i\beta$, $\beta = (4M^2 - 1)^{\frac{1}{2}}$.   Near $r = a$,

$$\xi \simeq |r - a|^{\frac{1}{2}} \cos(\tfrac{1}{2}\beta \ln|r-a| + \phi),$$

where ϕ is so chosen that the boundary conditions are satisfied.

Again, defining $\delta W = \delta W_1 + \delta W_2 + \delta W_3$ as in Case 1, and letting $\psi = \tfrac{1}{2}\beta \ln \epsilon + \phi$, we finally obtain

$$\delta W = \frac{a^3 \epsilon \, \xi_o^2 \, \alpha'^2}{4(m^2 + k^2 a^2)} \left[(4M^2 - 1)^{\frac{1}{2}} \sin 2\psi + (1 - 2M^2)(1 + \cos 2\psi) \right].$$

As $\epsilon \to 0$, $\psi \to -\infty$, so $\sin 2\psi$ and $\cos 2\psi$ take all values between -1 and $+1$. Hence the term in square brackets in δW takes all values between 1 and $1 - 4M^2$, since its turning value is at $\tan 2\psi = (4M^2 - 1)^{\frac{1}{2}}/(1 - 2M^2)$. Since $4M^2 > 1$, δW can be negative.

The conclusion is that the stability criterion for the particular perturbation considered by Suydam, and therefore a necessary condition for overall stability, is

$$M^2 \leqslant 1$$

$$\text{i.e.} \quad \frac{8\mu_o p_o'}{rB_z^2} \left(\frac{\mu}{\mu'}\right)^2 \geqslant -1,$$

which can be put in the form

$$\left(\frac{r\mu'}{\mu}\right)^2 + \frac{8r\mu_o p_o'}{B_z^2} \geqslant 0. \tag{3.115}$$

This must be satisfied at every value of r, and is termed a local stability criterion.

We discussed briefly the physical meaning of $\mu = B_\theta/rB_z$, earlier in the section. We see from (3.115) that, since the second term must be predominantly negative for confined plasma, the first term plays a crucial role, and to be non-zero requires $\mu' \neq 0$. Since μ' is a measure of the shear of adjacent magnetic field lines, a general conclusion is that a high degree of shear is required for stability. This shear prevents the simple interchange of flux tubes which we considered in §3-6-1, in particular the special case of interchange of flux tubes of equal magnetic flux.

3-6-5 The Newcomb criterion[10]

The Suydam criterion has been particularly useful in application, but one must remember its incompleteness. It is clear that a necessary and sufficient criterion is really what is required if a costly experiment is to be designed on the basis of stability theory. Such a criterion for cylindrical systems does exist,

65

but is different from the Suydam criterion in one essential
respect.   The Newcomb criterion, which is necessary and sufficient
for stability, is a non-local criterion only expressible as an
abstract mathematical theorem.   It will be given here without proof
for completeness, simply because in this short monograph it is
impossible to do justice to it, and to give a shortened proof
lacking rigour would detract from its theoretical beauty, but the
interested reader is strongly recommended to read Newcomb's original
paper.

The theorem can be stated thus:

A necessary and sufficient condition for the stability of a
cylindrical plasma is that (i) the solutions of the Euler equation
of the energy integral which are small at the singular points
shall have no zeros between the singular points;   (ii) Euler
solutions which satisfy the physical boundary condition at $r = 0$,

$$\xi(0) = 0 \qquad\qquad |m| \neq 1$$
$$\xi'(0) = 0 \qquad\qquad |m| = 1$$

shall have no zeros between $r = 0$ and the nearest singular point.
(iii)   A similar requirement applies in the interval (r_n, R),
where r_n is the outermost singular point.

A singular point is such that ξ'' in the Euler equation becomes
infinite.   As we saw in the previous section, this occurs where
$\alpha = 0$, that is, $m B_\theta/r + kB_z = 0$.

Near a singular point $r = a$, an Euler solution will behave as
$\xi_0(r - a)^\nu$, where ν is given by (3.114).   When Suydam's criterion
is satisfied the two roots are real, say $-n_1$ and $-n_2$ and such that
$n_1 + n_2 = 1$.   Taking $n_2 > n_1$, $(r - a)^{-n_2}$ is necessarily infinite
at $r = a$, but $(r - a)^{-n_1}$ may be zero since n_1 may be negative.
In any case, the "small" Euler solution at the singular point
$r = a$ is defined to be the one which behaves as $(r - a)^{-n_1}$.

3-7   EFFECT OF FINITE TRANSPORT COEFFICIENTS

In §3-2 we gave a very brief account of one particular effect of
finite electrical resistivity, the diffusion of magnetic flux
through the plasma.   In this short section, we shall review some
of the other consequences of finite transport coefficients.

(i) Electrical resistivity

Suydam's stability criterion (§3-6-4) is a necessary condition
that perturbations do not grow which are highly localised in the

neighbourhood of a point where $mB\theta/r + kB_z = 0$. The assumption of
zero resistivity is an essential ingredient of Suydam's theory and
therefore it is important to investigate the consequences of finite
resistivity. The results are two-fold. Firstly, it is found
that finite resistivity allows instabilities to develop when ideal
magnetofluid theory predicts stability. Secondly, other finite
resistivity instabilities are predicted. One such new instability
is known as the 'tearing' mode. Consider the situation where a
slab of plasma separates two regions of magnetic field directed in
opposite senses. In ideal magnetofluid theory, the system is
stable. However, finite resistivity allows diffusion of the field
across the plasma and eventually a break-up of the plasma slab and
reconnection of the magnetic field into a new configuration of
lower total energy. Figs. 3.9 and 3.10 illustrate the instability.

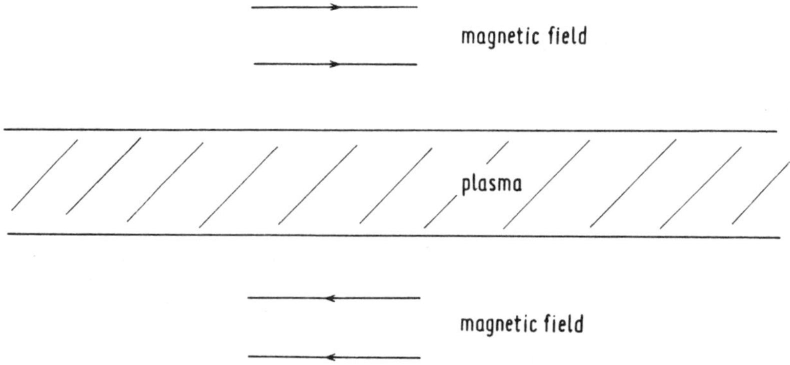

Fig. 3.9 Stable equilibrium (ideal magnetofluid theory)

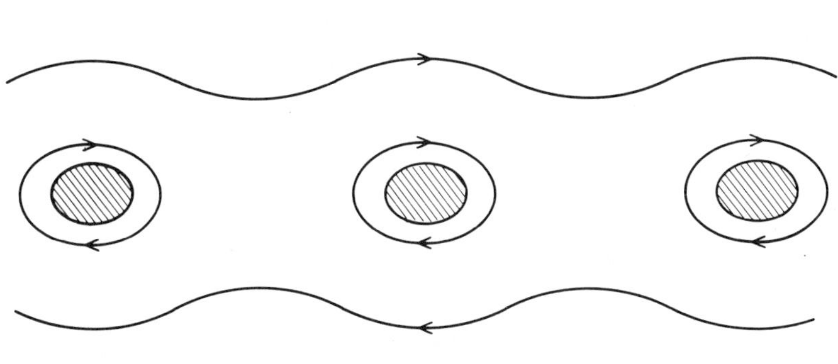

Fig. 3.10 Tearing mode (finite resistivity)

(ii) Anisotropic transport coefficients

It should be clear that in the presence of a strong magnetic field the motion of ions and electrons will be considerably modified, the effects being dependent not only on the local intensity of the magnetic field, but also on its direction. Essentially, electrons and ions are much freer to travel along the magnetic field than across it. The result is that transport properties will similarly depend on the magnitude and direction of the magnetic field, and for sufficiently high temperature and low density plasmas, transport coefficients appropriate to directions perpendicular and parallel to the magnetic field may differ by several orders of magnitude. For example, the parallel thermal conductivity is much greater than its perpendicular counterpart, while the parallel electrical resistivity is much less than the perpendicular resistivity.

3-8 ANISOTROPIC PRESSURE

One effect of anisotropy in transport coefficients is that the assumption of scalar pressure can no longer be maintained.
Thus, even if we can validly neglect effects of finite transport processes, we should still anticipate that the presence of a strong magnetic field will so influence particle motion that pressure will not be isotropic. In general therefore, a scalar pressure will be replaced by a tensor pressure $\underline{\underline{p}} = (p_{ij})$. The effect, for example, in the magnetofluid equation of motion (3.7), is that the term ∇p is replaced by $\nabla \cdot \underline{\underline{p}}$.

A considerable simplicity in the structure of the pressure is obtained, when we note that we should expect local cylindrical symmetry about the direction of the magnetic field. As a result, the pressure $\underline{\underline{p}}$ is locally diagonalised, if we transform to coordinates parallel and perpendicular to the magnetic field. We can express this either in the form

$$\underline{\underline{p}} = p_\perp \underline{\underline{1}} + (p_\parallel - p_\perp)\, \underline{\underline{bb}}, \tag{3.116}$$

where $\underline{\underline{1}}$ is the unit tensor and $\underline{b}$ is a unit vector parallel to the magnetic field; or

$$\underline{\underline{p}} = \begin{pmatrix} p_\perp & 0 & 0 \\ 0 & p_\perp & 0 \\ 0 & 0 & p_\parallel \end{pmatrix}$$

In the case of scalar pressure, we usually invoke the ideal gas equation of state $p = n\kappa T$, or for adiabatic motion $\frac{d}{dt}\left(p\rho^{-\gamma}\right) = 0.$ We may enquire what replaces this in the case of the pressure tensor above.   Chew, Low and Goldberger[11] showed that in a certain approximation, essentially the neglect of heat transfer along the magnetic field, the pressure components $p_{\parallel}$ and $p_{\perp}$ separately satisfy the adiabatic equations

$$\frac{d}{dt}\left(\frac{p_{\parallel}\, B^2}{\rho^3}\right) = 0$$

$$\frac{d}{dt}\left(\frac{p_{\perp}}{B\rho}\right) = 0. \tag{3.117}$$

A consequence of the double adiabatic approximation, as (3.117) is called, is the appearance of new magnetofluid instabilities. We shall simply quote results here (cf. Tayler).[12]   Qualitatively, instabilities occur if one of the pressure components is greatly in excess of the other.   The <u>stability</u> criteria are given below:

$$\frac{p_{\perp}^2}{6\left(p_{\perp} + \frac{B^2}{2\mu_o}\right)} < p_{\parallel} < p_{\perp} + \frac{B^2}{\mu_o}. \tag{3.118}$$

In (3.118), the right hand inequality is described as the 'firehose instability', because of its similarly to the well-known phenomenon which occurs if water is pumped at too high a rate of flow through a flexible hose.

Chapter 4

PLASMA KINETIC THEORY

In Chapter 3, we assumed without any profound physical justification that a plasma, an ensemble of electrons and ions which can sustain high frequency oscillations, can be adequately described in terms of a magnetofluid model. In this chapter one of our objectives is to give a theoretical justification of the fluid model, and in that sense this chapter should precede Chapter 3 in logical order. However, there are many situations in which a statistical treatment of plasma behaviour at the microscopic level is required, which goes far beyond a fluid model. Eventually, of course, we must derive from the theory experimentally measurable quantities such as the density, pressure and temperature.

In the first part of the chapter we introduce the concept of a kinetic equation, which describes the evolution of the distribution function of the system. The Boltzmann and Fokker-Planck equations are given a plausible derivation, and it is shown in general that the lower velocity moments of these equations lead to a set of equations closely similar in structure to the fluid equations of Chapter 3, though with an essential difference which poses a formidable mathematical problem.

The Vlasov equation is then obtained as a limiting case in which collisions are (apparently) neglected, and its general validity discussed. We study the important phenomenon of Landau damping, and apply Nyquist techniques to the case of unstable distributions. The presence of a uniform magnetic field in a plasma is also considered.

In the last part of the chapter we delve more deeply into non-equilibrium statistical mechanics and describe briefly the classical N-body problem, with Liouville's equation as the fundamental starting point. We obtain a more logical derivation of the Vlasov equation and give a brief review of recent attempts to derive a kinetic equation appropriate for describing the relaxation of a plasma to thermodynamic equilibrium, again with the Liouville equation as starting point.

4-1 KINETIC EQUATIONS

We introduce the concept of a kinetic equation as a description
of the statistical behaviour of a system of identical interacting
particles as it evolves in time. We first define a distribution
$f(\underline{r}, \underline{v}, t)$. Thus, let $f(\underline{r}, \underline{v}, t)\, d\underline{r}\, d\underline{v}$ be the number of particles
at time t with velocities in the range $(\underline{v}, \underline{v} + d\underline{v})$ in the
infinitesimal region $(\underline{r}, \underline{r} + d\underline{r})$. By integrating the distribution
function over all velocities, we obtain the particle density:

$$n\,(\underline{r},\,t)\ =\ \int d\underline{v}\ f(\underline{r},\,\underline{v},\,t).$$

This is the zero-th velocity moment, and we can carry this procedure further to obtain other macroscopic quantities.

$$\text{Mean velocity}\ =\ \underline{u}\ =\ \frac{1}{n}\int \underline{v}\ f\ d\underline{v}$$

$$\text{Pressure tensor}\ =\ \underline{\underline{p}}\ =\ m\int (\underline{v} - \underline{u})(\underline{v} - \underline{u})\ f\ d\underline{v}$$

$$\text{Energy density}\ =\ \varepsilon\ =\ \frac{\tfrac{1}{2}m}{n}\int (\underline{v} - \underline{u})^2\ f\ d\underline{v}$$

$$\text{Heat flux vector}\ =\ \underline{q}\ =\ \tfrac{1}{2}m\int (\underline{v} - \underline{u})^2\ (\underline{v} - \underline{u})\ f\ d\underline{v}.$$

We can formally set up an equation for f, the kinetic equation,
as follows,

$$\frac{\partial f}{\partial t} + \underline{v} \cdot \frac{\partial f}{\partial \underline{r}} + \underline{a} \cdot \frac{\partial f}{\partial \underline{v}}\ =\ \left(\frac{\partial f}{\partial t}\right)_c. \tag{4.1}$$

The LHS, with $\underline{a} = \underline{K}/m$ = acceleration due to an external force $\underline{K}$, is
just the total rate of change with time, of $f(\underline{r}, \underline{v}, t)$ namely,

$$\frac{df}{dt}\ =\ \frac{\partial f}{\partial t} + \frac{\partial f}{\partial \underline{r}} \cdot \frac{d\underline{r}}{dt} + \frac{\partial f}{\partial \underline{v}} \cdot \frac{d\underline{v}}{dt}.$$

The RHS is the time rate of change of f due to collisions;
although its detailed properties are not known at this stage, we
shall suppose conservation of particle number, total momentum and
total energy. The last two quantities would not be conserved
for a multi-species system, in particular a plasma of ions and
electrons, for in that case there would be a distribution function
for each species. Momentum and energy would then be conserved
only for the entire system. We shall not take up this point
here. It is evident that a straightforward extension of the
results will be available.

In terms of the general collision term, these conservation laws
may be expressed in the form

$$\int Q \left(\frac{\partial f}{\partial t}\right)_c d\underline{v} = 0,$$

where $Q = 1$, $m\underline{v}$, $\tfrac{1}{2}mv^2$. By forming $\int Q \, d\underline{v}$ of the kinetic equation
(4.1) we then obtain a corresponding set of equations which do
not involve details of the interparticle interaction.   These
equations depend only on $\underline{r}$ and t, and have a simple physical
interpretation at the macroscopic or fluid level.   We shall
suppose that the acceleration term $\underline{a}$ in (4.1) is either independ-
ent of $\underline{v}$ or assumes the form of the magnetic acceleration $\frac{e}{m} \underline{v} \times \underline{B}$.

Particle number

$$\int dv \left[\frac{\partial f}{\partial t} + \underline{v} \cdot \frac{\partial f}{\partial \underline{r}} + \underline{a} \cdot \frac{\partial f}{\partial \underline{v}}\right] = 0,$$

$$\text{i.e.} \quad \frac{\partial n}{\partial t} + \frac{\partial}{\partial \underline{r}} \cdot (n\underline{u}) = 0 \qquad (4.2)$$

for either form of $\underline{a}$.

Equation (4.2) is just the equation of continuity.

Momentum

$$\int d\underline{v} \left[m\underline{v} \frac{\partial f}{\partial t} + m\underline{v}\underline{v} \cdot \frac{\partial f}{\partial \underline{r}} + m\underline{v}\underline{a} \cdot \frac{\partial f}{\partial \underline{v}}\right] = 0.$$

We analyse the integral term by term, noting that

$$\int d\underline{v} \, m\underline{v} \frac{\partial f}{\partial t} = \frac{\partial}{\partial t} (nm\underline{u})$$

$$\int d\underline{v} \, m\underline{v}\underline{v}f = \int d\underline{v} \left[(\underline{v} - \underline{u})(\underline{v} - \underline{u}) + \underline{u}\underline{u} + 2\underline{u}(\underline{v} - \underline{u})\right] mf$$

$$= \underline{\underline{p}} + mn\underline{u}\underline{u}$$

$$\int d\underline{v} \, m\underline{v}\underline{a} \cdot \frac{\partial f}{\partial \underline{v}} = -mn\underline{a}, \text{ if } \underline{a} \text{ is independent of } \underline{v}$$

$$= -en \, \underline{u} \times \underline{B}, \text{ for the magnetic acceleration.}$$

The two forms of the momentum equation are thus

$$\rho \left(\frac{\partial \underline{u}}{\partial t} + \underline{u} \cdot \frac{\partial \underline{u}}{\partial \underline{r}}\right) = -\frac{\partial}{\partial \underline{r}} \cdot \underline{\underline{p}} + \underline{A}, \qquad (4.3)$$

where $\underline{A} = \rho \underline{a}$ or $ne \, \underline{u} \times \underline{B}$, respectively, and (4.2) has been used in
deriving the equation.   Since the current density $\underline{j} = ne\underline{u}$, the
second form of $\underline{A}$ is the familiar term $\underline{j} \times \underline{B}$.

<u>Energy</u>

$$\int d\underline{v} \left[\tfrac{1}{2}m\underline{v}^2 \, \frac{\partial f}{\partial t} + \tfrac{1}{2}m\underline{v}^2 \, \underline{v} \cdot \frac{\partial f}{\partial \underline{r}} + \tfrac{1}{2}m\underline{v}^2 \, \underline{a} \cdot \frac{\partial \overline{f}}{\partial \underline{v}} \right] = 0.$$

Following the same procedure as before, we now obtain an equation
for the energy density ε.

The resulting equation is

$$\frac{\partial}{\partial t}(n\varepsilon) + \frac{\partial}{\partial \underline{r}} \cdot (n\varepsilon\underline{u}) = -\frac{\partial}{\partial \underline{r}} \cdot \underline{q} - \underline{\underline{p}} : \frac{\partial \underline{u}}{\partial \underline{r}} \cdot$$

$$(4.4)$$

There is no contribution to the energy equation for either form of
$\underline{a}$; (4.4) takes the form given above after incorporating (4.2) and
(4.3). The terms on the RHS are, respectively, the flow of heat,
and the work done due to changes in the shape and volume of the
plasma element. A kinetic temperature T $(\underline{r}, t)$ may be defined by
the relation

$$\frac{3}{2}\kappa T = \varepsilon,$$

and an alternative form to (4.4) is

$$\frac{3}{2} n\kappa \left(\frac{\partial T}{\partial t} + \underline{u} \cdot \frac{\partial T}{\partial \underline{r}} \right) = -\frac{\partial}{\partial \underline{r}} \cdot \underline{q} - \underline{\underline{p}} : \frac{\partial \underline{u}}{\partial \underline{r}} \cdot$$

An important point to note is that, although resemblance to the
standard equations of fluid mechanics is undoubtedly real, equations
(4.2)-(4.4) do not form a closed set of partial differential
equations. So far they represent no more than a valuable set of
relations which must hold between the various macroscopic quantities.
To see this, let us define the quantity $\int \underline{v}^n f \, d\underline{v}$ in general as an
nth order moment of f. Then, (4.2) is a relation between zero-th
and first order moments, (4.3) between first and second, (4.4)
between second and third. We have no defining equations for the
fourth order moments, or higher. We are thus led to the conclusion
that we require details of the collision term and in fact a
solution of the kinetic equation in order to obtain the macroscopic properties of the system. Nevertheless, the fluid concept
has proved extremely useful in practice; it may be that an
approximate, and even poor, solution of the kinetic equation yields
a reliable set of fluid equations.

In order to apply these ideas to a plasma of ions and electrons,
we shall have two distribution functions, f_i and f_e, and require
two kinetic equations, one for each species. Then, we obtain two
distinct equations of continuity, since the number of electrons and

ions is individually conserved. However, since momentum and
energy are conserved only for the entire ensemble of electrons and
ions, we obtain, as we should expect, a single equation of motion
and energy equation. Further, the current density and charge
density

$$\underline{j} = e \int \underline{v} \left(f_i - f_e \right) d\underline{v}$$

$$\text{and} \quad q = e \int \left(f_i - f_e \right) d\underline{v}$$

are sources in Maxwell's equations. We shall come back to these
points again later.

4-1-1 Boltzmann equation

Our starting point is the general form for any kinetic equation,
(4.1). The collision term $\left(\dfrac{\partial f}{\partial t} \right)_c$ so far has been perfectly general,
satisfying only general conservation laws. The reason for writing
it in the particular form above is to indicate that it is the rate
at which the distribution function changes, at a particular value
of $\underline{r}$, $\underline{v}$ and t, solely because of particle collisions. Thus, the
collision term measures the difference between the rate at which
particles with velocity $\underline{v}' \neq \underline{v}$ are scattered into the range
$(\underline{v}, \underline{v} + d\underline{v})$ and the rate at which particles are scattered out of
the same range of velocities. One particular form of the collision
term is due to Boltzmann.

Consider a collision between a particle of velocity $\underline{v}$ and one of
velocity $\underline{u}$. Suppose that after the collision their velocities are
$\underline{v}'$ and $\underline{u}'$. We shall consider only elastic collisions. Let $\underline{G}$
be the velocity of the centre of mass of the pair of particles.
Since $\underline{G}$ is conserved, and noting that we are dealing with a system
of identical particles,

$$\underline{G} = \tfrac{1}{2}(\underline{u} + \underline{v}) = \tfrac{1}{2}(\underline{u}' + \underline{v}').$$

Let $\underline{g}$, $\underline{g}'$ be their relative velocity before and after colliding.
Then

$$\underline{g} = \underline{u} - \underline{v}, \qquad \underline{g}' = \underline{u}' - \underline{v}'.$$

Since total energy is conserved, $u^2 + v^2 = u'^2 + v'^2$. Hence
$g = g'$.

A complete specification of the collision requires knowledge of
the scattering angle θ and the plane of collision ϕ. Consider now

collisions which occur in a volume element $d\underline{r}$ in a time dt between particles in the range $(\underline{v}, \underline{v} + d\underline{v})$ and $(\underline{u}, \underline{u} + d\underline{u})$ respectively, with θ in the range $d\theta$ and ϕ in the range $d\phi$.   Provided that we can neglect correlations between particles when outside $d\underline{r}$, the number of collisions is proportional to:

(i)   number of particles in $d\underline{v}\ d\underline{r}\ =\ f\ (\underline{r},\ \underline{v},\ t)\ d\underline{v}\ d\underline{r}$

(ii)   number of particles in $d\underline{u}$ which are in the segment $d\phi$ of the cylinder of unit cross-sectional area and length gdt.

(iii) $\sigma(g,\theta)\ \sin\theta\ d\theta$, σ being the differential scattering cross section, which we assume to depend only on relative velocities and the scattering angle.

Hence, the number of collisions leading to particles being scattered <u>out of</u> $d\underline{v}$ is

$$f(\underline{r},\ \underline{v},\ t)\ f(\underline{r},\ \underline{u},\ t)\ g\sigma(g,\ \theta)\ \sin\theta\ d\theta\ d\phi\ d\underline{v}\ d\underline{u}\ d\underline{r}\ dt.$$

In these collisions, the final velocities $\underline{u}'$ and $\underline{v}'$ lie in the range $d\underline{u}'$, $d\underline{v}'$ respectively and $\underline{u}'$, $\underline{v}'$ depend on $\underline{u}$, $\underline{v}$, θ, ϕ through the law of interaction.

Now we consider the inverse situation.   Particles initially have velocities $\underline{u}'$, $\underline{v}'$ and finally, $\underline{u}$, $\underline{v}$.   Hence the number of inverse collisions is given by

$$f(\underline{r},\ \underline{v}',\ t)\ f(\underline{r},\ \underline{u}',\ t)\ g'\sigma(g',\ \theta)\ \sin\theta\ d\theta\ d\phi\ d\underline{v}'\ d\underline{u}'\ d\underline{r}\ dt.$$

But $g' = g$ and $d\underline{g}' = d\underline{g}$, since $\underline{g}' = R(\underline{g})$, a rotation of the vector $\underline{g}$ without change of magnitude.   Further, by transforming from the pair of variables $\underline{v}$, $\underline{u}$ to $\underline{G}$, $\underline{g}$ and noting that the Jacobian of the transformation is unity, we obtain

$$d\underline{v}\ d\underline{u}\ =\ \left|\frac{\partial(\underline{v},\ \underline{u})}{\partial(\underline{G},\ \underline{g})}\right|\ d\underline{G}\ d\underline{g}\ =\ d\underline{G}\ d\underline{g},$$

and similarly

$$d\underline{v}'\ d\underline{u}'\ =\ \left|\frac{\partial(\underline{v}',\ \underline{u}')}{\partial(\underline{G},\ \underline{g}')}\right|\ d\underline{G}\ d\underline{g}'\ =\ d\underline{G}\ d\underline{g}'\ =\ d\underline{G}\ d\underline{g}.$$

Hence, $d\underline{v}\ d\underline{u} = d\underline{v}'\ d\underline{u}'$, and we obtain for the number of particles scattered <u>into</u> $d\underline{v}$

$$f(\underline{r},\ \underline{v}',\ t)\ f(\underline{r},\ \underline{u}',\ t)\ g\sigma\ (g,\ \theta)\ \sin\theta\ d\theta\ d\phi\ d\underline{v}\ d\underline{u}\ d\underline{r}\ dt.$$

Thus the net number of particles scattered into $d\underline{v}$ is

$$d\underline{v}\ d\underline{r}\ dt \int d\underline{u}\ d\Omega\ g\sigma(g,\ \theta)\ \{f(\underline{v}')\ f(\underline{u}') - f(\underline{v})\ f(\underline{u})\},$$

and this is to be identified with

$$d\underline{v}\ d\underline{r}\ dt\ \left(\frac{\partial f}{\partial t}\right)_c \ ,$$

so that the resulting kinetic equation is

$$\frac{\partial f}{\partial t} + \underline{v}\cdot\frac{\partial f}{\partial \underline{r}} + \underline{a}\cdot\frac{\partial f}{\partial \underline{v}} = \int d\underline{u}\ d\Omega\ g\sigma(g,\ \theta)\ \{f(\underline{v}')\ f(\underline{u}') - f(\underline{v})\ f(\underline{u})\}.$$

$$(4.5)$$

We shall not in this book discuss applications and methods of solution of the Boltzmann equation (4.5).   The interested reader is referred to the Bibliography for further discussion on this topic.

Several points should be noted, many of which apply to any kinetic equation.

1. It is not at all obvious that a system of N particles can be described in terms of a single-particle distribution function.

2. The form of the collision term is not a consequence of dynamical laws, though it is consistent with them.

3. We have treated $d\underline{r}$ and $d\underline{v}$ in the derivation as infinitesimal quantities.   However, they must not be too small or the statistical nature by which we obtained the number of collisions will be meaningless.   As a consequence, in practice we are forced to work, not with f, but more correctly with

$$\bar{f} = \frac{1}{\Delta\underline{r}\ \Delta\underline{v}}\int_{\Delta\underline{r}\ \Delta\underline{v}} f\ d\underline{r}\ d\underline{v}.$$

In effect, the system must be 'coarse-grained'.

4. Only binary collisions are considered, and for high densities, and also for long-range forces, deviations must be expected due to multiple particle interactions.

5. The form of the Boltzmann collision integral, involving the products $f(\underline{u})\,f(\underline{v})$ and $f(\underline{u}')\ f(\underline{v}')$ implies not just binary collisions but complete absence of correlations between colliding particles. This will be taken up again later.

4-1-2 Fokker-Planck equation

In some respects this is the other extreme in kinetic equations. We now abandon the Boltzmann assumption of binary collisions.   In this approach, we single out a single particle as a marker, or test particle, and observe its motion as it interacts with a large number of particles.   These multi-particle collisions independently scatter the test particle in a random manner.

We omit any spatial variation in the theoretical description
which follows. We shall adopt the basic ideas of Markov processes
in statistical theory. Thus, let $f(\underline{v}, t)$ be the velocity distribution of the system of particles at time t. We study the
evolution of the system by obtaining a relation between the distri-
bution $f(\underline{v}, t + \Delta t)$ and $f(\underline{v}, t)$. To this end, let us postulate
the existence of a transition probability $Q(\underline{v} - \Delta \underline{v}, \Delta \underline{v})$ which gives
the probability that the test particle suffers a large number of
collisions during the time interval Δt resulting in its velocity
changing from $\underline{v} - \Delta \underline{v}$ to $\underline{v}$. We have to take Δt long enough for a
large number of independent collisions to take place, but short
enough for the net change $\Delta \underline{v}$ to be small. This imposes some
constraint on the particle density and the strength of the interaction. We then assert that the required relation is

$$f(\underline{v}, t + \Delta t) \;=\; \int d(\Delta \underline{v}) \; f(\underline{v} - \Delta \underline{v}, t) \; Q(\underline{v} - \Delta \underline{v}, \Delta \underline{v}).$$

It should be emphasised again that this is an assumption which is
being made, not a statement which can be given a mathematical proof.

We now define the following averages:

$$\langle \Delta \underline{v} \rangle \;=\; \int d(\Delta \underline{v}) \; \Delta \underline{v} \; Q(\underline{v}, \Delta \underline{v})$$

$$\langle \Delta \underline{v} \, \Delta \underline{v} \rangle \;=\; \int d(\Delta \underline{v}) \; \Delta \underline{v} \; \Delta \underline{v} \; Q(\underline{v}, \Delta \underline{v})$$

and expand f and Q to second order in a Taylor series about $\Delta \underline{v} = 0$.
Thus,

$$f(\underline{v}, t + \Delta t) \;=\; f(\underline{v}, t) - \frac{\partial}{\partial \underline{v}} \cdot \left[\langle \Delta \underline{v} \rangle \, f \right] + \tfrac{1}{2} \frac{\partial^2}{\partial \underline{v} \, \partial \underline{v}} : \left[\langle \Delta \underline{v} \, \Delta \underline{v} \rangle \, f \right].$$

For sufficiently small Δt, we obtain the rate of change of f due to
collisions, so the collision term in the kinetic equation (4.1) is
in this description given by

$$\left(\frac{\partial f}{\partial t} \right)_c \;=\; - \frac{\partial}{\partial \underline{v}} \cdot (\underline{D}_1 \, f) + \tfrac{1}{2} \frac{\partial^2}{\partial \underline{v} \, \partial \underline{v}} : (\underline{\underline{D}}_2 \, f), \tag{4.6}$$

$$\text{where} \quad \underline{D}_1 \;=\; \lim_{\Delta t \to 0} \frac{\langle \Delta \underline{v} \rangle}{\Delta t}$$

$$\underline{\underline{D}}_2 \;=\; \lim_{\Delta t \to 0} \frac{\langle \Delta \underline{v} \, \Delta \underline{v} \rangle}{\Delta t}$$

$\underline{D}_1$ is known as the coefficient of dynamical friction, $\underline{\underline{D}}_2$ the

dispersion or velocity diffusion coefficient. They can be
determined once Q is known.

4-1-3 Relation between Boltzmann and Fokker-Planck equations

The Boltzmann and Fokker-Planck equations have been derived under
quite different physical assumptions, and have different mathematical form. Yet in certain situations the Boltzmann equation
exhibits the basic structure of a Fokker-Planck collision term
(4.6).

We start with the Boltzmann collision integral, involving the
binary collision of two particles with initial and final velocities
$\underline{u}$, $\underline{v}$ and $\underline{u}'$, $\underline{v}'$ respectively. Let us now suppose that the
collision is weak, so that $\underline{u}' = \underline{u} - \frac{1}{2}\Delta\underline{v}$, $\underline{v}' = \underline{v} + \frac{1}{2}\Delta\underline{v}$, chosen to
conserve momentum. Then, in the Boltzmann equation,

$$\left(\frac{\partial f}{\partial t}\right)_c = \int d\underline{u}\ d\Omega\ g\sigma\ (g,\ \theta)\left[f(\underline{v} + \tfrac{1}{2}\Delta\underline{v})\ f(\underline{u} - \tfrac{1}{2}\Delta\underline{v}) - f(\underline{v})\ f(\underline{u})\right].$$

We can now consider $\Delta v \ll u$, v and expand. The zero order terms
cancel, and we retain only up to second order. It is evident that
the collision term above must take the general form given in (4.6).
The main difference is that $\underline{D}_1$ and $\underline{\underline{D}}_2$ are determined not just in
terms of $\sigma(g,\ \theta)$, which is a property only of particle interactions,
but also in terms of the distribution function f, which is strictly
unknown until the Boltzmann equation has been solved. Nevertheless,
in practice an estimate of $\underline{D}_1$ and $\underline{\underline{D}}_2$ is obtained from binary
collisions, suitably averaged over a Maxwellian velocity distri-
bution, and the resulting equation is solved in the linear approx-
imation, neglecting the dependence of $\underline{D}_1$ and $\underline{\underline{D}}_2$ on f.

4-2 VLASOV EQUATION

The Vlasov equation has been one of the most important tools of
plasma kinetic theory in understanding collective processes which
go on in a plasma. It is not even a kinetic equation, in that it
can describe only reversible phenomena, and these only for times
short compared with the collision relaxation time. A simple-
minded derivation of the Vlasov equation is to set the collision
term to zero in the kinetic equation, thus obtaining the following
equations, for ions and electrons,

$$\frac{\partial f_i}{\partial t} + \underline{v}\cdot\frac{\partial f_i}{\partial \underline{r}} + \frac{e}{m_i}\ (\underline{E} + \underline{v}\times\underline{B})\cdot\frac{\partial f_i}{\partial \underline{v}} = 0 \qquad (4.7)$$

$$\frac{\partial f_e}{\partial t} + \underline{v} \cdot \frac{\partial f_e}{\partial \underline{r}} - \frac{e}{m_e} (\underline{E} + \underline{v} \times \underline{B}) \cdot \frac{\partial f_e}{\partial \underline{v}} = 0. \quad (4.8)$$

Strictly, $\underline{E}$ and $\underline{B}$ are, at this stage, externally applied fields.
An argument is then usually given which amounts to self-consistency.
Thus, we first assume only external fields in which the plasma
particles move, then apply Maxwell's equations in which the charge
and current densities are given as in §4-1, namely,

$$\underline{j} = e \int \underline{v} \, (f_i - f_e) \, d\underline{v}$$

$$q = e \int (f_i - f_e) \, d\underline{v}.$$

In the literature, the approximation is often made of treating a
plasma as a one-component system, namely, an electron plasma, in
which only the electrons play a dynamical rôle. The more massive
ions merely serve to provide a positive neutralising background.
However, situations arise in which ion dynamics is dominant. This
is not a difficulty in principle, and the techniques for handling
the electron plasma can usually be readily extended to include ion
motion.

We end this introductory discussion on the Vlasov equation with
some dimensional arguments about time-scales which may help to
clarify ideas about criteria for validity of the Vlasov equation.

The parameters which characterise the behaviour of an electron
plasma can only be combinations of the fundamental quantities e,
m, n, β, respectively the charge and mass of an electron, the
electron density and the inverse thermal energy $\beta = \dfrac{1}{\kappa T}$. From
these we can construct a non-dimensional parameter (setting
$\varepsilon_o = 1$ for the present)

$$\Gamma = e^2 \, n^{\frac{1}{3}} \, \beta.$$

For a typical high temperature plasma, $\Gamma \ll 1$. Physically, $e^2 \, n^{\frac{1}{3}}$
is the average electrostatic potential energy between a pair of
electrons, since $n^{-\frac{1}{3}}$ is their average separation distance, while $\dfrac{1}{\beta}$
is the average kinetic energy of an electron in a plasma at
temperature T. Hence, Γ can be interpreted as the ratio (average
potential energy)/(average kinetic energy), and as can be verified
is typically very small. Γ is sometimes termed the diluteness
parameter. We can now form all characteristic times for a plasma

with the expression

$$\tau = (m\beta)^{\frac{1}{2}} \, n^{-\frac{1}{3}} \, \Gamma^{y}, \qquad y \text{ arbitrary,}$$

though only two important cases arise, with $y = -\frac{1}{2}$ and $y = -2$.

(i) $\quad y = -\frac{1}{2}. \quad \tau = \left(\dfrac{m}{e^{2}n}\right)^{\frac{1}{2}} \simeq \tau_{p}$, the plasma period

(ii) $\quad y = -2 \quad \tau = (e^{4}n)^{-1}\left(\dfrac{m}{\beta^{3}}\right)^{\frac{1}{2}} \simeq \tau_{r}$, the relaxation time.

τ_{p} is just the inverse of the plasma frequency ω_{p} introduced in §1.2.2. It is the only characteristic time independent of β. τ_{r} is obtained by noting the general property of any relaxation time, that it should be inversely proportional to the particle density. Further, $\dfrac{\tau_{p}}{\tau_{r}} = \Gamma^{\frac{3}{2}} \ll 1$, so that we expect two widely separated time scales. We shall put this to a different use later in discussion of a more fundamental approach to kinetic theory. Here we use the result above to underline the possibility that for times much longer than τ_{p} the Vlasov equation will still give a very useful description of plasma effects, and only when τ_{r} is approached should we doubt our conclusions.

The question of reversibility was briefly mentioned at the beginning of this section. Let it suffice at this stage that if an equation is unchanged by the substitution $t \rightarrow -t$, $\underline{v} \rightarrow -\underline{v}$, then it is reversible. This is obviously the case for the Vlasov equation, but not for the Boltzmann and Fokker-Planck equations because of the collision terms. The deeper implications of irreversibility and the "time-arrow" are discussed at length in treatises on statistical mechanics, and we shall not attempt more than to make appropriate comments on this subject when the need arises.

4-2-1 Landau damping

Landau[13] solved the classical problem of electrostatic oscillations of an unmagnetised electron plasma, described by a spatially uniform time-independent velocity distribution $f_{o}(\underline{v})$. He considered in detail the case where f_{o} is Maxwellian.

We have thus to solve the following system of equations for the electron distribution f:

$$\frac{\partial f}{\partial t} + \underline{v} . \nabla f - \frac{e}{m} \underline{E} . \frac{\partial f}{\partial \underline{v}} = 0$$

$$\epsilon_o \nabla \cdot \underline{E} = e \left[n_o - \int f \, d\underline{v} \right], \qquad (4.9)$$

where n_o is the uniform ion background density.

We consider only the case of small amplitude oscillations, so that $f = f_o(\underline{v}) + f_1(\underline{r}, \underline{v}, t)$, f_1 being a small perturbation imposed on f_o, of the same order of magnitude as the electric field $\underline{E}$. Next, we introduce an electrostatic potential ϕ, so that $\underline{E} = -\nabla \phi$, and the equations (4.9) reduce, in first order, to the form

$$\frac{\partial f_1}{\partial t} + \underline{v} \cdot \nabla f_1 + \frac{e}{m} \nabla \phi \cdot \frac{\partial f_o}{\partial \underline{v}} = 0$$

$$\epsilon_o \nabla^2 \phi = e \int f_1 \, d\underline{v}, \qquad (4.10)$$

where we have used $\int f_o d\underline{v} = n_o$.

Since f_o is a function only of $\underline{v}$, and equations (4.10) are linear in f_1 and ϕ, the technique of Fourier analysis suggests itself. Thus, if we assume the forms

$$f_1 = \sum_{\underline{k}, \omega} f_1(\underline{v}, \underline{k}, \omega) \exp(i\underline{k} \cdot \underline{r} - i\omega t)$$

and a similar expression for ϕ, except for velocity dependence

$$\phi = \sum_{\underline{k}, \omega} \phi(\underline{k}, \omega) \exp(i\underline{k} \cdot \underline{r} - i\omega t),$$

we expect to obtain, as a result of satisfying (4.10), a dispersion relation of the form $F(\underline{k}, \omega) = 0$. This in fact was the procedure adopted by Vlasov, who first attempted to solve the problem. However, his results involved a divergent integral, which he tried to avoid by taking a principal value.

Landau obtained a correct solution by restating the problem thus:

Assume that a definite non-equilibrium f_1 exists initially. The problem is to determine the resulting motion of the plasma.

This restatement is not trivial, and suggests that, as an initial value problem, we should analyse the time dependence by the technique of Laplace transforms. Fourier analysis of the spatial dependence of f_1 and ϕ is satisfactory.

We first perform the spatial Fourier analysis, obtaining the counterpart of (4.10) for $f_1(\underline{k}, \underline{v}, t)$ and $\phi(\underline{k}, t)$:

$$\frac{\partial f_1}{\partial t} + i\underline{k} \cdot \underline{v} f_1 + i \frac{e}{m} \phi \, \underline{k} \cdot \frac{\partial f_o}{\partial \underline{v}} = 0$$

$$\varepsilon_o k^2 \phi = -e \int f_1 \, d\underline{v}. \qquad (4.11)$$

It may be useful here to remind the reader of the elements of Laplace transforms. The Laplace transform of a function $g(t)$ is defined by

$$\bar{g}(p) = \int_0^\infty g(t) \, e^{-pt} \, dt, \quad \mathrm{Re}(p) > 0$$

and the inverse transform by

$$g(t) = \frac{1}{2\pi i} \int_C \bar{g}(p) \, e^{pt} \, dp,$$

where the p-integration is to be carried out along the straight line C from $p = -i\infty + \sigma$ to $p = i\infty + \sigma$, and σ is a positive constant sufficiently large that C lies to the right of all singularities of $\bar{g}(p)$. C, together with the infinite semi-circle enclosing the left-hand half-plane of p, is sometimes referred to as the Bromwich contour in texts on functions of a complex variable and contour integration.

Applying this to (4.11), we have to multiply through by e^{-pt} and integrate over t. Since

$$\int_0^\infty \frac{\partial f_1}{\partial t} \, e^{-pt} \, dt = f_1 e^{-pt} \Big|_0^\infty + p \int_0^\infty f_1 \, e^{-pt} \, dt$$

$$= -\psi + p \, \bar{f}_1 \, (\underline{k}, \, \underline{v}, \, p),$$

where $\psi = f_1 (\underline{k}, \, \underline{v}, \, t = 0)$,

equations (4.11) become

$$(p + i\underline{k}.\underline{v}) \, \bar{f}_1 + i \frac{e}{m} \, \bar{\phi} \underline{k}. \, \frac{\partial f_o}{\partial \underline{v}} = \psi$$

$$\varepsilon_o k^2 \bar{\phi} = -e \int \bar{f}_1 \, d\underline{v}. \qquad (4.12)$$

Hence,

$$\bar{f}_1 = (p + i\underline{k}.\underline{v})^{-1} \left[\psi - i \frac{e}{m} \bar{\phi} \, \underline{k}. \, \frac{\partial f_o}{\partial \underline{v}} \right] \qquad (4.13)$$

and substituting into the second equation of (4.12) gives

$$\bar{\phi} = - \frac{e}{\varepsilon_o k^2} \frac{\displaystyle \int \frac{\psi(\underline{v}) \, d\underline{v}}{p + i\underline{k}.\underline{v}}}{\displaystyle 1 - \frac{ie^2}{\varepsilon_o m k^2} \int \frac{\underline{k}. \, \frac{\partial f_o}{\partial \underline{v}} \, d\underline{v}}{p + i\underline{k}.\underline{v}}} \qquad (4.14)$$

The electron distribution and electrostatic potential are now in
principle determined, given an arbitrary initial distribution ψ.
We now particularise to the case of a Maxwellian distribution

$$f_o(\underline{v}) = n_o \left(\frac{m}{2\pi\kappa T}\right)^{\frac{3}{2}} \exp\left(-\frac{mv^2}{2\kappa T}\right) .$$

In evaluating $\bar{\phi}$ using (4.14) the x-axis can be chosen arbitrarily
to lie along $\underline{k}$.   Letting $v_x = u$,

$$\psi(u) = \int \psi(\underline{v}) \, dv_y \, dv_z$$

$$\text{and} \quad f_o(u) = n_o \left(\frac{m}{2\pi\kappa T}\right)^{\frac{1}{2}} \exp\left(-\frac{mu^2}{2\kappa T}\right)$$

$$\bar{\phi} = -\frac{e}{\varepsilon_o k^2} \frac{\displaystyle\int_{-\infty}^{\infty} \frac{\psi(u)}{p + iku} \, du}{1 - \frac{ie^2}{\varepsilon_o mk} \displaystyle\int \frac{df_o(u)}{du} \frac{du}{p + iku}} . \quad (4.15)$$

$\bar{\phi}$, considered as a function of the complex variable p, is defined
only in the right hand half-plane, $\text{Re}(p) > 0$.   However, we can
define $\bar{\phi}$ in the left hand half-plane as the analytical continuation
of equation (4.15).   Now, if $\psi(u)$ is an entire function of u, that
is, has no singularities in the finite u-plane, then the integral
$\int_{-\infty}^{\infty} \frac{\psi(u) \, du}{p + iku}$, continued analytically to the left hand half-plane of
p, also defines an entire function of p.   In practice, the analyt-
ical continuation is defined in a slightly modified **form**, thus:

$$\text{Re}(p) > 0 \quad \int_{-\infty}^{\infty} \frac{\psi(u) \, du}{p + iku} .$$

The path of integration is along the real u-axis

$$\text{Re}(p) < 0 \quad \int_{\Gamma} \frac{\psi(u) \, du}{p + iku} .$$

The path of integration is along the real axis, except for the
singularity at $u = -\frac{p}{ik}$, as shown below.

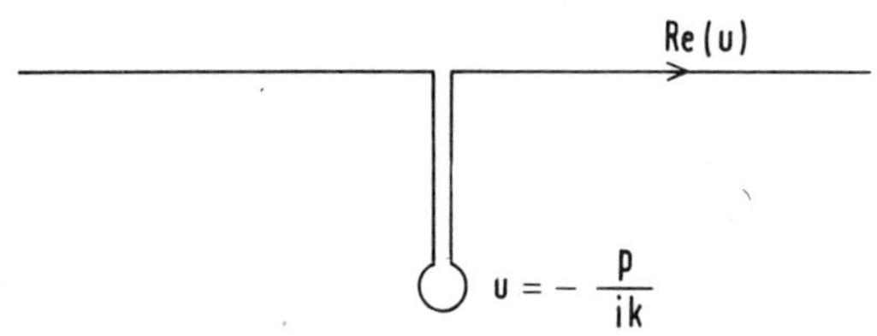

This function has no singularities at finite values of p.

Since $\dfrac{df_o}{du}$ is also an entire function of u, the integral in the denominator of (4.15) can be treated in a similar manner. Hence, $\bar{\phi}$ is the ratio of two entire functions, so that its singularities occur only at the zeros of the denominator, and then only in the left hand half-plane, since when $\text{Re}(p) > 0$ the denominator must always have a non-vanishing imaginary part. We can therefore determine the asymptotic form, for large t, of $\phi(t)$:

$$\phi(t) \;=\; \frac{1}{2\pi i} \int_{-i\infty + \sigma}^{i\infty + \sigma} \bar{\phi}(p)\; e^{pt}\; dp.$$

Because of the properties of $\bar{\phi}$, the path of integration can be displaced to the left, going around all the poles of $\bar{\phi}$ it meets. Let p_k denote the first pole it encounters, that is, the pole with the smallest (negative) real part. The original and displaced paths are shown below.

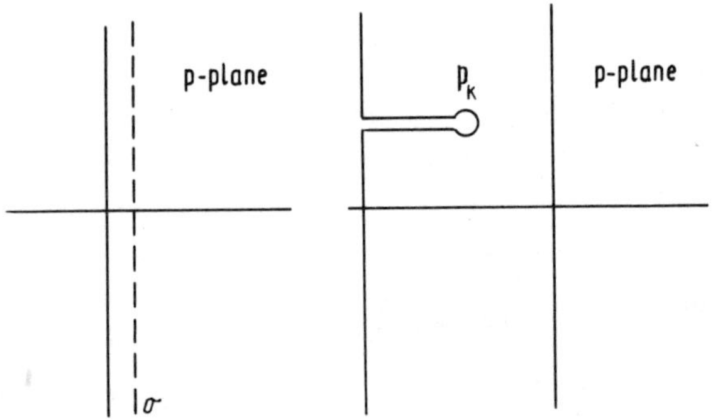

p_k satisfies the relation

$$\frac{ie^2}{\varepsilon_o mk} \int_\Gamma \frac{df_o}{du} \; \frac{du}{p + iku} \;=\; 1. \tag{4.16}$$

Since we are considering large t, only the residue at p_k will be significant in the integral. Thus, for large t, $\phi(t) \simeq \exp(p_k t)$. Since p_k is possibly complex but has a negative real part, it follows that $\phi(t)$ is exponentially damped in time and is possibly periodic. It is very important to realise that this damping phenomenon occurs only in $\phi(t)$. The perturbed distribution behaves quite differently, as can be seen from (4.13). The asymptotic form of f_1, after the effect of $\bar{\phi}$ has disappeared, is simply

$$f_1(t) \;=\; \frac{1}{2\pi i} \int_{-i\infty + \sigma}^{i\infty + \sigma} \frac{\psi(v)}{p + iku}\; e^{pt}\; dp,$$

which is purely oscillatory, with behaviour e^{-ikut}. This must be
the case, since the Vlasov equation is reversible.

Before proceeding with a more detailed look at the properties
of Landau damping, a few general remarks may prove useful.

1. The existence or absence of Landau damping is determined both
   by the nature of $f_o(\underline{v})$ and the initial perturbation $\psi(\underline{v})$.
   Damping results from roots of the dispersion relation (4.16) in
   the lower half-plane of u. For a Maxwellian, these are the
   only roots. For other forms of $f_o(\underline{v})$, as we shall shortly see,
   there may be roots also in the upper half-plane, which will lead
   to unstable modes.

2. The model of Landau damping described above may be criticised on
   the grounds that it is a linear theory, but there is no doubt
   that such a theory contains Landau damping. In fact, in
   an interesting set of experiments, Malmberg and Wharton[14] have
   established without doubt that Landau damping of small amplitude electrostatic waves occurs, in appropriate conditions,
   and obtain quantitative agreement. For large amplitude waves,
   they observe a certain degree of particle trapping. This is
   in accord, qualitatively, with results of non-linear theory,
   but any further discussion of non-linear effects is unfortunately
   beyond the scope of this book.

3. Much has been written on the physical interpretation of Landau
   damping, and we shall outline briefly the main contenders, with
   comments.

   (i) An early candidate for providing an explanation was 'phase-
       mixing', namely, that if many small amplitude waves of a
       whole range of frequencies are superimposed, then they will
       ultimately average out to zero. This 'explanation' was
       discounted in the Malmberg experiments, which showed that
       electrons of only a very narrow range of velocities, centred
       around the phase velocity $\frac{\omega}{k}$ of the wave, contributed to
       damping.

   (ii) The most popular contender has arisen from the observation
        that Landau damping is proportional to the slope of the
        distribution function at the phase velocity of the wave (in
        accord with experiment). If we have a distribution reversed
        in slope at the phase velocity we should expect energy to be
        fed into the wave and instead of damping we should expect to

find an unstable mode. This argument depends for success
on the possibility that an electron moving slightly _slower_
than the electrostatic wave is accelerated by the wave and
gains energy, while one moving slightly faster gives up
energy since it suffers a deceleration. There is in fact a
flaw in this argument since strictly we are dealing with a
small-amplitude phenomenon. All that happens is that
electrons speed up and slow down alternately. To obtain
the effect above requires a degree of particle trapping.

(iii) Stix[15] gives an excellent account of Landau damping. Only
a brief comment, without mathematical details, will be given
here.

He considers the problem of the absorption of energy by a
group of charged particles drifting through the electric field
of a plasma wave. When a large number of particles have
velocities slightly less than the phase velocity of the wave,
and somewhat fewer particles have velocities slightly faster,
there is a net absorption of energy from the electric field.
This is the case when the electron distribution function has
positive gradient at the phase velocity, and leads to damping
of the wave, that is, the phenomenon of Landau damping. The
reverse, namely growth of the wave amplitude, occurs with a
negative slope.

The main difference between Stix's description and that given
by others, is that he obtains a quantitatively correct physical picture. Application of elementary perturbation theory
yields a first order correction to the velocity of an electron
due to the presence of a small-amplitude electric field. The
rate of change of particle energy is then calculated to
second order and the result averaged over all electrons, giving
the correct expression for Landau damping.

The dispersion relation, (4.16) can only be solved numerically
for arbitrary wave number k. It is instructive to look at two
special cases for which an analytic solution is possible.

(i) Long wavelength perturbations, $k \to 0$.

Here, the point $u = -\dfrac{p}{ik}$ is displaced to very large $|u|$. Hence,
since $f_o(u)$ decreases rapidly as $|u| \to \infty$, we shall obtain a good
approximation by integrating along Γ only along the real axis and
omitting the contribution from the pole. We can therefore expand
the factor $(p + iku)^{-1}$ in powers of k. Thus,

$$\frac{1}{k} \int_\Gamma \frac{df_o}{du} \; \frac{du}{p + iku} \;\; \simeq \;\; \frac{1}{k} \int_{-\infty}^{\infty} du \; \frac{df_o}{du} \; \frac{1}{p} \left(1 - \frac{ikm}{p} - \frac{k^2 u^2}{p^2} \cdots \right) .$$

Then, using the fact that $\displaystyle\int_{-\infty}^{\infty} \frac{df_o}{du} \, du = f_o \Big|_{-\infty}^{\infty} = 0$, the leading term of the dispersion relation is

$$\frac{e^2}{\varepsilon_o m p^2} \int_{-\infty}^{\infty} u \, \frac{df_o}{du} \, du \;\; = \;\; 1,$$

$$\text{giving} \quad \frac{e^2 n_o}{\varepsilon_o m p^2} \;\; = \;\; -1,$$

i.e. $\quad p_k = -i\omega_o, \quad \omega_o = \left(\dfrac{n_o e^2}{\varepsilon_o m} \right)^{\frac{1}{2}} = $ plasma frequency.

The next term in the expansion leads to

$$p_k = -i\omega_o \left(1 + \frac{3}{2} \lambda^2 k^2 \right),$$

where λ is the Debye length, $(\varepsilon_o \kappa T / n_o e^2)^{\frac{1}{2}}$.

Note that there must actually be a very small damping term, since we have neglected that part of the integration along Γ which leads to a negative $\mathrm{Re}(p)$. It is not a difficult exercise, and is shown in Landau's original paper, that if $p_k = -i\omega - \gamma$ then to a good approximation

$$\gamma \;\; = \;\; \omega_o \left(\frac{\pi}{8} \right)^{\frac{1}{2}} x^3 \exp(-\tfrac{1}{2}x^2), \quad x = \frac{1}{k\lambda} .$$

The result is valid provided $\gamma \ll \omega_o$, implying that $k\lambda \ll 1$.

(ii) Short wavelength perturbations, $k \to \infty$.

Again referring to Landau, the result now is that ω and γ, using the same definition $p = -i\omega - \gamma$, both tend to infinity with k, but such that, for k large, $\omega \ll \gamma$, $\dfrac{\omega}{k} \to 0$ and $\dfrac{\gamma}{k} \to \infty$.

The precise form of the solution is unimportant here. The main point is that short wavelength perturbations are strongly damped and are of high frequency, while long wavelength perturbations, agreeing with elementary theory, oscillate at the plasma frequency and suffer only very little damping, especially when the wavelength is much greater than a Debye length.

In the general case of arbitrary k, we can write the dispersion relation in the form

$$\frac{\omega_p^2}{\pi^{\frac{1}{2}} k^2 \alpha^2} \int_\Gamma \frac{-2\zeta\, e^{-\zeta^2}\, d\zeta}{\zeta - ip} = 1,$$

where $\alpha^2 = \dfrac{2\kappa T}{m}$, $\zeta = u/\alpha$.

Thus $\quad k^2 \lambda^2 + 1 + i\pi^{\frac{1}{2}}\, zW(z) = 0$,

where $\quad z = ip/k\alpha$

and $\quad W(z) = \dfrac{i}{\pi} \int_\Gamma \dfrac{\exp(-\zeta^2)\, d\zeta}{z - \zeta}$.

$W(z)$ has been tabulated, and is known as the plasma dispersion function.[16]

4-2-2 Stability and Nyquist diagrams

Our starting point in this section is the dispersion relation (4.16)

$$\frac{ie^2}{\varepsilon_o mk} \int_\Gamma \frac{df_o}{du} \frac{du}{p + iku} = 1,$$

where now f_o is any arbitrary function of velocity. We shall develop a method of investigating the stability of the plasma with respect to electrostatic waves.

We first show that all single-maximum distributions, of which the Maxwellian is a representative, have only time decaying roots, with $\mathrm{Re}(p) < 0$. Let us assume initially that a pole does exist with $\mathrm{Re}(p) > 0$. This implies that Γ is just the real axis, so that

$$1 = \frac{e^2}{\varepsilon_o mk} \int_{-\infty}^{\infty} \frac{df_o}{du} \frac{du}{y - ix + ku}, \quad \text{where } p = x + iy \text{ and } x > 0.$$

Hence, equating real and imaginary parts,

$$1 = \frac{e^2}{\varepsilon_o mk} \int_{-\infty}^{\infty} du\, \frac{df_o}{du} \frac{ku + y}{(ku + y)^2 + x^2}$$

$$0 = \frac{e^2 x}{\varepsilon_o mk} \int_{-\infty}^{\infty} \frac{df_o}{du} \frac{du}{(ku + y)^2 + x^2} . \tag{4.17}$$

Suppose now that f_o has its single maximum at $u = U$, so that $\dfrac{df_o}{du} \lessgtr 0$ according as $u \gtrless U$. Then, in the second relation of (4.17), the denominator is positive definite. Now multiply this relation by $\dfrac{(kU + y)}{x}$ and subtract from the first relation. The result is

$$1 = -\frac{e^2}{\varepsilon_o mk} \int_{-\infty}^{\infty} (U - u) \frac{df_o}{du} \frac{du}{(ku + y)^2 + x^2} .$$

Since $(U - u) \frac{df_o}{du} \geqslant 0$ everywhere, RHS < 0, which is impossible. Hence there can be no poles with $Re(p) > 0$, which proves our result. Note that this argument cannot be used if $x < 0$, since then Γ is not along the real axis of u. However, this can give only decaying terms.

We now give two simple illustrations of instabilities.

(i) The first was discussed by Buneman,[17] who considered the case of two well-defined interpenetrating beams. In this case the dispersion relation takes on the purely algebraic form

$$\frac{\omega_1^2}{(\omega + kv_1)^2} + \frac{\omega_2^2}{(\omega + kv_2)^2} = 1,$$

where $\omega_{1,2}$ are the plasma frequencies of the two beams, and $v_{1,2}$ are the beam velocities.

As k increases, at first four real roots occur, $\omega \simeq 0$ (twice) and $\omega \simeq \pm (\omega_1^2 + \omega_2^2)^{\frac{1}{2}}$. Eventually, two roots become complex, one the complex conjugate of the other, and the root of the form $\omega = \omega_r - i\gamma$ corresponds to exponential growth.

If the beams are identical, $\omega_1 = \omega_2 = \omega_o$, $v_1 = -v_2 = u$, we get

$$\omega_o^2 \left[\frac{1}{(\omega - ku)^2} + \frac{1}{(\omega + ku)^2} \right] = 1,$$

with roots

$$\omega^2 = \tfrac{1}{2}\omega_o^2 \left[1 + 2 \left(\frac{ku}{\omega_o}\right)^2 \pm \sqrt{1 + 8 \left(\frac{ku}{\omega_o}\right)^2} \right]$$

which can become negative.

(ii) The example given above is trivial in that there is no thermal spread in the beams, and the results can be obtained without recourse to the methods of this chapter. The second illustration shows the effect of thermal spread on the stability of inter-penetrating beams. Consider therefore two interpenetrating Maxwellian plasmas, with relative velocity V. For example, we could take

$$f_o(u) = \exp\left[-\left(\frac{(u - \tfrac{1}{2}V)}{\alpha}\right)^2\right] + \exp\left[-\left(\frac{(u + \tfrac{1}{2}V)}{\alpha}\right)^2\right] .$$

We know from the last section that single-maximum distributions are always stable. Hence a necessary condition for $f_o(u)$ above to

be <u>unstable</u> is that it has a minimum at all.   This condition is
$V > \sqrt{2}\alpha$.   This is not a sufficient instability criterion.   A
numerical study[17] based on the method given below leads to the
criterion $V > 1.85\alpha$.

We now outline a general method of analysing the stability of
a system, given $f_o(u)$.   We shall write the dispersion relation in
the following form, setting $p = iz$,

$$1 - \frac{e^2}{\varepsilon_o mk} \int_\Gamma \frac{df_o}{du} \frac{du}{ku + z} = 0.$$

Then the system is stable if there are no solutions z for given k
with negative imaginary parts.

Define

$$F(k, z) = 1 - \frac{e^2}{\varepsilon_o mk} \int_\Gamma \frac{df_o}{du} \frac{du}{ku + z}. \qquad (4.18)$$

The stability criterion is that $F(k, z)$ shall have no zeros in the
lower half z-plane.   The only singularities of F lie on the real
axis.

As $k \to \infty$, $F(k, z) \to 1$ uniformly in z, off the real axis.

As $|z| \to \infty$, $F(k, z) \to 1$ uniformly in k, $\arg(z) \neq 0, \pi$.
Thus, for sufficiently large k, F has no zeros and the plasma is
stable against modes of sufficiently large k.

The number of zeros of F in the lower half plane may be investigated by contour integration methods.   Since F is analytic in
the lower half plane, the number of zeros there is given by

$$n(k) = \frac{1}{2\pi i} \int_C \frac{\frac{\partial F}{\partial z}}{F} \, dz, \qquad (4.19)$$

where C is a contour (followed anticlockwise) of a large semi-
circle in the lower half plane and the straight line $z = \omega - i\eta$,
with ω real and $\eta \to +0$, that is, the line just below the real axis.
Now let us transform to the new complex variable

$$\zeta = F(k, z).$$

Then,

lower half z-plane $\to$ domain D of ζ-plane

contour C $\to$ contour $\gamma(k)$ within D.

Hence, $$n(k) = \frac{1}{2\pi i} \int_{\gamma(k)} \frac{d\zeta}{\zeta}$$

= number of times origin $\zeta = 0$ is encircled by $\gamma(k)$.

90

This is known as the Nyquist method, first used in investigating the properties of electrical circuits, and $\gamma(k)$ is sometimes termed a Nyquist diagram.

We can get a different form of the stability criterion. Suppose the system is <u>unstable</u> for some k'. Then $\gamma(k)$ must encircle the origin at least once. But there exists a value of k, say k", so large that $\gamma(k")$ does not encircle the origin. This follows from the first property of F above. However, since F is continuous in k, $\gamma(k)$ is a continuously changing contour as a function of k, so that there exists an intermediate value of k, say k''', such that $\gamma(k''')$ passes through the origin. We now use the second property of F, namely, $F \to 1$ as $|z| \to \infty$ to show that the semi-circle in C $\to$ point (1, 0) in $\gamma(k)$. Hence, $\zeta = 0$ on $\gamma(k''')$ must correspond to some point $\omega - i\eta$, $\eta \to +0$, in the z-plane. Thus, if the system is unstable, the quantity

$$\varepsilon(k''', \omega) = \lim_{\eta \to +0} F(k''', \omega - i\eta) = 0$$

for some value of ω.

Conversely, if $\varepsilon(k, \omega)$ vanishes anywhere for real k, ω the system is unstable.

The quantity $\varepsilon(k, \omega)$, as defined above, is the generalised dielectric coefficient of a plasma. To see this, let us recall the elementary derivation of ε for a cold plasma given in §1-2-3. We now seek the response of the plasma to an electric field $\underline{E} = \underline{E}_o \exp i(\omega t + \underline{k} \cdot \underline{r})$, producing a small perturbation f_1 in the distribution function, so that $f = f_o(\underline{v}) + f_1(\underline{v}) \exp i(\omega t + \underline{k} \cdot \underline{r})$. The linearised Vlasov equation is

$$\frac{\partial f_1}{\partial t} + \underline{v} \cdot \frac{\partial f_1}{\partial \underline{r}} - \frac{e\underline{E}}{m} \cdot \frac{\partial f_o}{\partial \underline{v}} = 0.$$

Substitution of the form above for $f_1(\underline{r}, \underline{v}, t)$ gives

$$f_1(\underline{v}) = f_1(\underline{k}, \underline{v}, \omega) = \frac{\frac{e}{m}\underline{E} \cdot \frac{\partial f_o}{\partial \underline{v}}}{i(\omega + \underline{k} \cdot \underline{v})},$$

noting again the singularity at $\omega + \underline{k} \cdot \underline{v} = 0$. In this problem, the electric field is imposed from outside and we could have followed the Landau method of treating it as an initial value problem. An alternative point of view is to suppose that the field is switched on adiabatically at $t = -\infty$, since it is an external field. Thus, suppose $\underline{E} \simeq e^{\eta t}$, η small and positive, for $t < 0$. We have then to replace the denominator by $i(\omega + \underline{k}.\underline{v}) + \eta$. In the limit $\eta \to +0$, the response of the plasma, in the form of an induced current density $\underline{j}$, is

$$\underline{j} = \lim_{\eta \to +0} - \frac{e^2}{m} \underline{E} \cdot \int \frac{\frac{\partial f_o}{\partial \underline{v}} \underline{v} \, d\underline{v}}{i(\omega + \underline{k}.\underline{v}) + \eta} \, .$$

Inserting this expression for $\underline{j}$ into Maxwell's equations gives an identification of the dielectric coefficient $\varepsilon(\underline{k}, \omega)$ with the previous form

$$\varepsilon(\underline{k}, \omega) = \lim_{\eta \to +0} F(\underline{k}, \omega - i\eta) \, .$$

4-2-3 Dielectric properties of a uniformly magnetised plasma

To study the properties of a plasma in a non-uniform magnetic field, which is the subject of primary interest in laboratory plasmas, would take us too far afield in this introductory book. The case of a uniform magnetic field is sufficiently tractable, and contains sufficient similarities with the general problem, that we shall devote this section to one important aspect, the dielectric coefficient of a hot plasma.

We have already obtained, in §1-2-3, the basic structure of the dielectric coefficient in the case of a cold plasma in a uniform magnetic field. As we shall see, this is unaltered in form. We follow the classic paper by Bernstein[18] in the discussion.

The Vlasov equation now includes a term involving the uniform magnetic field $\underline{B}$.

$$\frac{\partial f}{\partial t} + \underline{v} \cdot \frac{\partial f}{\partial \underline{r}} - \frac{e}{m} (\underline{E} + \underline{v} \times \underline{B}) \cdot \frac{\partial f}{\partial \underline{v}} = 0. \qquad (4.20)$$

We consider only the response of the plasma to a small electrostatic field $\underline{E}$, and no perturbed magnetic fields are considered in our treatment, which follows closely the basic ideas of the last section, §4-2-2. Initially, then, $\underline{E} = 0$ and $f = f_o(\underline{v})$. We now switch on a small field $\underline{E}$ and determine the small change f_1, $f = f_o + f_1$. The appropriate linearised equation for f_1 is

$$\frac{\partial f_1}{\partial t} + \underline{v} \cdot \frac{\partial f_1}{\partial \underline{r}} - \frac{e}{m} \underline{v} \times \underline{B} \cdot \frac{\partial f_1}{\partial \underline{v}} = \frac{e}{m} \underline{E} \cdot \frac{\partial f_o}{\partial \underline{v}} \, . \qquad (4.21)$$

Singular integrals are to be defined following Landau's procedure, which allows us to carry out a Fourier analysis of both space and time variations. (The alternative procedure of adiabatic

switching at $t = -\infty$, as in §4-2-2, gives identical results.)
Hence

$$i(\omega + \underline{k}.\underline{v})\, f_1 - \frac{e}{m}\, \underline{v} \times \underline{B} \cdot \frac{\partial f_1}{\partial \underline{v}} = \frac{e}{m}\, \underline{E} \cdot \frac{\partial f_o}{\partial \underline{v}} \,. \tag{4.22}$$

Method of solution

We introduce a cylindrical coordinate system for the velocity
variable, with axis parallel to $\underline{B}$, along the unit vector $\underline{b}$.   Thus,

$$\underline{v} = v_\parallel\, \underline{b} + \underline{v}_\perp, \quad \underline{v}_\perp = v_\perp\,(\cos\phi,\ \sin\phi)$$

$$\underline{B} = B\underline{b}.$$

Similarly, let $\underline{k} = k_\parallel\, \underline{b} + \underline{k}_\perp$ and take the polar axis $\phi = 0$ along
$\underline{k}_\perp$, $\underline{k}_\perp = k_\perp\,(1,\ 0)$.   Then (4.22) becomes, with $\Omega = \dfrac{eB}{m}$,

$$i(\omega + k_\parallel\, v_\parallel + k_\perp\, v_\perp \cos\phi)\, f_1 + \Omega\, \frac{\partial f_1}{\partial \phi} = \frac{e}{m}\, \underline{E} \cdot \frac{\partial f_o}{\partial \underline{v}}. \tag{4.23}$$

Restricting our discussion to the class of equilibria possessing
cylindrical symmetry about $\underline{b}$, which we should expect on physical
grounds, f_o is of form $f_o(v_\parallel,\ v_\perp)$.   Then, if $\underline{E} = (E_\parallel,\ E_{\perp 1},\ E_{\perp 2})$
the equation for f_1 is

$$i(\omega + k_\parallel\, v_\parallel + k_\perp\, v_\perp \cos\phi)\, f_1 + \Omega\, \frac{\partial f_1}{\partial \phi}$$

$$= \frac{e}{m}\left[E_\parallel \frac{\partial f_o}{\partial v_\parallel} + (E_{\perp 1} \cos\phi + E_{\perp 2} \sin\phi)\, \frac{\partial f_o}{\partial v_\perp} \right], \tag{4.24}$$

which is of form

$$i(a + b \cos\phi)\, f_1 + \frac{\partial f_1}{\partial \phi} = A + B \cos\phi + C \sin\phi \tag{4.25}$$

with suitably defined coefficients a, b, A, B, C.   Thus

$$\frac{\partial}{\partial \phi}\, f_1 \exp i(a\phi + b \sin\phi)$$

$$= (A + B \cos\phi + C \sin\phi)\, \exp i(a\phi + b \sin\phi).$$

Hence,

$$f_1 = \exp -i(a\phi + b \sin\phi)\ \times$$

$$\times \left[\int^\phi d\phi'\, (A + B \cos\phi' + C \sin\phi')\, \exp i(a\phi' + b \sin\phi') + \text{constant} \right].$$

We now use the series expansion

$$\exp ib \sin\phi' = \sum_n J_n\,(b)\, \exp in\phi'$$

and obtain

$$f_1 = \exp -i(a\phi + b\sin\phi) \times$$

$$\times \left\{ \sum_n J_n(b) \left[A \frac{\exp\ i(a+n)\phi}{i(a+n)} + \frac{B}{2}\left(\frac{\exp\ i(a+n+1)\phi}{i(a+n+1)} + \right.\right.\right.$$

$$\left.\left. + \frac{\exp\ i(a+n-1)\phi}{i(a+n-1)}\right) + \frac{C}{2i}\left(\frac{\exp\ i(a+n+1)\phi}{i(a+n+1)} - \frac{\exp\ i(a+n-1)\phi}{i(a+n-1)}\right)\right] +$$

$$\left. + \text{constant} \right\}. \qquad (4.26)$$

To evaluate the constant in expression (4.26), we note that $f_1(\phi + 2\pi) = f_1(\phi)$ for continuity. Hence constant is zero, and we have

$$f_1 = -i \sum_n J_n(b) \exp\ i(n\phi - b\sin\phi) \times$$

$$\times \left\{ \frac{A}{a+n} + \frac{B}{2}\left(\frac{e^{i\phi}}{a+n+1} + \frac{e^{-i\phi}}{a+n-1}\right) + \frac{C}{2i}\left(\frac{e^{i\phi}}{a+n+1} - \frac{e^{-i\phi}}{a+n-1}\right)\right\}.$$

To form $\underline{j}$, the induced current density, we have to multiply f_1 by $-e\underline{v}$ and integrate over $\underline{v}$, using $d\underline{v} = v_\perp\, dv_\parallel\, dv_\perp\, d\phi$. The integrations are all straightforward and we shall simply state results.

$$j_\alpha = -\frac{ie^2 n_o}{m} \sum_{n\beta\gamma} \int_{-\infty}^{\infty} dv_\parallel \int_0^\infty v_\perp dv_\perp\, A^n_{\alpha\beta}\, B^n_{\beta\gamma}\, E_\gamma, \qquad (4.28)$$

where

$$\left(B^n_{\beta\gamma}(x) \right) = \begin{bmatrix} \dfrac{1}{x} & 0 & 0 \\[2ex] 0 & \dfrac{x}{x^2 - \Omega^2} & \dfrac{-i\Omega}{x^2 - \Omega^2} \\[2ex] 0 & \dfrac{i\Omega}{x^2 - \Omega^2} & \dfrac{x}{x^2 - \Omega^2} \end{bmatrix} \qquad (4.29)$$

and $x = \omega + n\Omega + k_\parallel v_\parallel$.

$$A^n_{\alpha\beta}(y) = -\sum_m J_n(y) J_m(y) a^{nm}_{\alpha\beta}, \qquad (4.30)$$

where $y = k_\perp v_\perp/\Omega$
and $\left(a^{nm}_{\alpha\beta} \right)$

$$= \mu \begin{bmatrix} \left(\delta_{n,m} + \delta_{n,m+2} + \delta_{n,m-2}\right) & \dfrac{1}{2i}\left(\delta_{n,m+2} - \delta_{n,m-2}\right) & \delta_{n,m+1} + \delta_{n,m-1} \\[2ex] \dfrac{1}{2i}\left(\delta_{n,m+2} - \delta_{n,m-2}\right) & \left(\delta_{n,m} - \delta_{n,m+2} - \delta_{n,m-2}\right) & -i\left(\delta_{n,m+1} - \delta_{n,m-1}\right) \\[2ex] \delta_{n,m+1} + \delta_{n,m-1} & -i\left(\delta_{n,m+1} - \delta_{n,m-1}\right) & \dfrac{\nu}{\mu}\delta_{n,m} \end{bmatrix}$$

where $\mu = \frac{1}{2} v_\perp \frac{\partial f_o}{\partial v_\perp}$, $\quad \nu = v_\parallel \frac{\partial f_o}{\partial v_\parallel}$.

The dielectric tensor can be formed from the current density, in terms of the coefficients A and B defined in (4.29) and (4.30).

$$\varepsilon_{ik} = 1 - \frac{\omega_p^2}{\omega^2} \sum_{nj} \int v_\perp \, dv_\perp \, dv_\parallel \, A_{ij}^n \, B_{jk}^n . \qquad (4.31)$$

Integrals occurring in (4.31) have denominators of various types, namely x^{-1} and $(x^2 - \Omega^2)^{-1}$ with x defined above. Integrals over $v_\parallel$ are singular integrals to be treated again by the Landau prescription, extended to cover the second type. Thus

$$\int_L \frac{g(t) \, dt}{z-t} \equiv P \int \frac{g(t) \, dt}{z-t} - i\pi g(z)$$

$$\int_L \frac{\Omega g(t) \, dt}{(z-t)^2 - \Omega^2} \equiv P \int \frac{\Omega g(t+z) \, dt}{t^2 - \Omega^2} - \frac{i\pi}{2} \left[g(z-\Omega) - g(z+\Omega) \right]$$

$\int_L$ denoting a Landau integral.

The first type of integral above represents Landau damping. The second type is associated with energy passed to the electrons by a cyclotron resonance with the magnetic field.

An interesting consequence here is the appearance of new resonances at multiples of the gyrofrequency Ω.

Special case

If the transverse wavelength of the perturbation is much greater than the electron thermal Larmor radius, then we can assume $\frac{k_\perp v_\perp}{\Omega} \ll 1$. (Larmor radius for a particular electron with component of velocity $v_\perp$ is $\frac{v_\perp}{\Omega}$. The thermal Larmor radius is this quantity averaged over the distribution, so strictly we should use only the averaged quantity above, but the results will be unaltered.) This results in the argument of the Bessel functions occurring in (4.30) being very much less than unity. In the limit, $J_o(y) \to 1$ and $J_n(y) \to O(y^n)$. This leads to the simple result

$$j_\parallel = \frac{e^2}{m} \int_L v_\parallel E_\parallel \frac{\partial f_o}{\partial v_\parallel} \frac{dy}{i(\omega + k_\parallel v_\parallel)}$$

$$j_{\perp 1} = \frac{e^2}{m} \int_L v_\perp \frac{\partial f_o}{\partial v_\perp} \left[\frac{(\omega + k_\parallel v_\parallel) E_{\perp 1}}{(\omega + k_\parallel v_\parallel)^2 - \Omega^2} + \frac{i\Omega E_{\perp 2}}{(\omega + k_\parallel v_\parallel)^2 - \Omega^2} \right]$$

and a similar expression for $j_{\perp 2}$.

4-3 The Liouville equation and kinetic theory

The basic difference between a plasma and a neutral gas lies in the different interparticle interactions. We shall study in this section the methods developed for neutral gases and discuss to what extent they are applicable to a description of plasma phenomena.

Consider a gas of N particles in volume V. In order to avoid explicit mention of boundary conditions, we shall invoke the thermodynamic limit, that is, N and V both tend to infinity in such a way that the density $n = \dfrac{N}{V}$ remains finite and constant. Our fundamental starting point, for a classical system, is the Liouville equation. Let $F_N\,(\underline{q}_1,\ \underline{p}_1;\ \underline{q}_2,\ \underline{p}_2;\ \cdots\cdots;\ \underline{q}_N,\ \underline{p}_N;\ t)$ be the distribution function in the phase space of N particles. F_N is defined as the probability that, at time t, particle 1 is in the state $\underline{q}_1$, $\underline{p}_1$, particle 2 in $\underline{q}_2$, $\underline{p}_2$, and so on. We also take all N particles identical and F_N symmetric in the interchange of any pair of particles. Then Liouville's equation can be expressed in the form

$$\frac{\partial F_N}{\partial t} + \left[F_N,\, H \right] = 0, \tag{4.33}$$

where H is the N-particle Hamiltonian. For the case of an external potential ψ and two-body potentials ϕ,

$$H = \sum_{j=1}^{N} \left[\frac{1}{2m}\, \underline{p}_j^2 + \psi(\underline{q}_j) \right] + \sum_{i<j}^{N} \phi\left(|\underline{q}_i - \underline{q}_j|\right).$$

The use of a probability distribution function needs only a brief comment. In earlier sections, the single-particle distribution $f(\underline{r},\ \underline{v},\ t)$, integrated over all $\underline{r}$ and $\underline{v}$, gives the total number of particles in the system. In the present discussion it is more convenient to adopt the normalisation

$$\int F_N\,(\underline{X}_1,\ \cdots\cdots \underline{X}_N;\ t)\ \prod_{i=1}^{N} d\underline{X}_i = 1,$$

where we have now introduced the notation $\underline{X}_i = (\underline{q}_i,\ \underline{p}_i)$.

We can now define <u>reduced</u> distribution functions

$$F_s\,(\underline{X}_1 \cdots\cdots \underline{X}_s;\ t) = V^s \int F_N\, d\underline{X}_{s+1} \cdots\cdots d\underline{X}_n.$$

Integrating the Liouville equation over $\underline{X}_{s+1} \cdots\cdots \underline{X}_N$ we get, for all s, a hierarchy of N equations

$$\frac{\partial F_s}{\partial t} = \left[H_s,\, F_s \right] + \frac{N-s}{V} \int d\underline{X}_{s+1} \left[\sum_{i=1}^{s} \phi_{i,s+1},\, F_{s+1} \right]$$

$$s = 1, 2 \dots N - 1$$

$$\frac{\partial F_N}{\partial t} = \left[H, F_N \right]. \tag{4.34}$$

This is sometimes termed the BBGKY[19] hierarchy, in recognition of the work over the years, by Bogoliubov, Born, Green, Kirkwood and Yvon, in developing this field.

In (4.34), H_s is the Hamiltonian appropriate to a group of s identical interacting particles.

$$H_s = \sum_{i=1}^{s} \left[\frac{1}{2m} \underline{p}_i^2 + \psi(\underline{q}_i) \right] + \sum_{i<j}^{s} \phi_{i,j},$$

while $\phi_{i,j}$ is just a short-hand form for $\phi\left(|\underline{q}_i - \underline{q}_j|\right)$.

For $s = 1$, and taking the thermodynamic limit,

$$\frac{\partial F_1}{\partial t} + \frac{1}{m} \underline{p}_1 \cdot \frac{\partial F_1}{\partial \underline{q}_1} - \frac{\partial \psi(\underline{q}_1)}{\partial \underline{q}_1} \cdot \frac{\partial F_1}{\partial \underline{p}_1} = n \int d\underline{q}_2 \, d\underline{p}_2 \, \frac{\partial \phi_{1,2}}{\partial \underline{q}_1} \cdot \frac{\partial F_2}{\partial \underline{p}_1}. \tag{4.35}$$

F_1 is the probability of finding any one particle in the state $(\underline{q}_1, \underline{p}_1)$ irrespective of the state of the rest of the system. F_2 is the corresponding probability of finding a pair of particles in prescribed states.

Note that, like all other equations in the hierarchy (4.34), equation (4.35) is exact. In particular it has retained the property of Liouville's equation in being reversible in time, and thus cannot adequately describe relaxation to equilibrium. Further, we cannot solve for F_1 unless we know F_2, and so on up to F_N. Some additional ideas are necessary to "break the chain". In equation (4.35):

(i)   we must make some assumption or approximation to get F_2;

(ii)   we need a physical (or mathematical) condition to convert equations for F_1 and F_2 into irreversible equations.

There are several ways of achieving some of these ends; the Vlasov equation is a consequence of the simplest procedure. We suppose that the motion and position of a pair of particles to be uncorrelated. Then,

$$F_2(\underline{X}_1, \underline{X}_2; t) = F_1(\underline{X}_1; t) \, F_1(\underline{X}_2; t).$$

Let us neglect ψ, which plays no rôle here, and define

$$f(\underline{q}_1, \underline{p}_1, t) = n F_1.$$

Let $U(\underline{q}_1, t) = \int \phi\left(|\underline{q}_1 - \underline{q}_2|\right) f\left(\underline{q}_2, \underline{p}_2, t\right) d\underline{q}_2 \, d\underline{p}_2.$

Then f satisfies the equation

97

$$\frac{\partial f}{\partial t} + \frac{p_1}{m} \cdot \frac{\partial f}{\partial \underline{q}_1} - \frac{\partial U}{\partial \underline{q}_1} \cdot \frac{\partial f}{\partial \underline{p}_1} = 0.$$

U is the potential interaction of all other particles of the system
with the particle at $\underline{q}_1$. This is precisely the form of the Vlasov
equation. It shows in effect that it is not <u>collisionless</u>, but
correlationless.

4-3-1 Multiple time scales

Since this is meant to be an elementary introduction to plasma
physics, some of the topics raised in this chapter are more likely
to stimulate questions than to answer them, for there is no
intention of delving deeply into modern kinetic theory and non-
equilibrium statistical mechanics. One of the main purposes
of this chapter is just to reveal some of the general work going on
in the development of the subject.

Modern kinetic theory of neutral gases, such as that developed
by Bogoliubov,[20] makes use of the short range nature of the interaction, which allows the introduction of a collision time τ_o and a
mean free time between collisions τ_1 with $\tau_o << \tau_1$. It then makes
good sense to talk of binary collisions, since typically $\tau_o/\tau_1 \simeq$
10^{-4}, and we have a good basis for an expansion procedure. In a
plasma, such concepts are ambiguous. The long-range nature of the
Coulomb force means that many-particle correlations are important.
Nevertheless, as we saw in §4-2, there are two significantly
different characteristic times in a plasma, τ_p and τ_r, so that again
we should expect to be able to develop a perturbation theory in
terms of the small parameter τ_p/τ_r.

We shall discuss only one approach here, due to Frieman[21] and
Sandri,[22] and refer the reader to other attempts which lead
essentially to their results. The starting point is the hierarchy
of equations for the reduced distribution functions F_s, and the
existence of a sequence of time scales, $\tau_o, \tau_1 \ldots$. Their
treatment is general, in that it is equally applicable to widely
different classes of physical systems. It is essentially the
characteristic of $\tau_o << \tau_1$ which allows information to be lost at
the short time scale level and thus yields irreversible behaviour.
To determine the appropriate perturbation expansion parameter ε
to be used, we have to classify the system according to the nature
of the particle interaction. We shall investigate three

98

possibilities, weak coupling, short range and long range. The
long range interaction represents most closely plasma behaviour;
short range effects are typical of neutral gas behaviour. Weak
coupling contains elements applicable to both plasmas and neutral
gases, and is the simplest in mathematical description. We shall
therefore look in more detail at the case of weak coupling, in
order to establish the basic ideas involved. Using equations
(4.34) and (4.35), we write out explicitly the equations for F_1
and F_2, in the case where there is no external potential ψ:

$$\frac{\partial F_1}{\partial t} + \underline{v}_1 \cdot \frac{\partial F_1}{\partial \underline{r}_1} = \frac{n}{m} \int dr_2 \; dv_2 \; \frac{\partial \phi}{\partial \underline{r}_1} \cdot \frac{\partial F_2}{\partial \underline{v}_1} \tag{4.36}$$

$$\frac{\partial F_2}{\partial t} + \underline{v}_1 \cdot \frac{\partial F_2}{\partial \underline{r}_1} + \underline{v}_2 \cdot \frac{\partial F_2}{\partial \underline{r}_2} - \frac{1}{m} \frac{\partial \phi}{\partial \underline{r}_1} \cdot \frac{\partial F_2}{\partial \underline{v}_1} - \frac{1}{m} \frac{\partial \phi}{\partial \underline{r}_2} \cdot \frac{\partial F_2}{\partial \underline{v}_2}$$

$$= \frac{n}{m} \int dr_3 \; dv_3 \left[\frac{\partial \phi}{\partial \underline{r}_1} \cdot \frac{\partial F_3}{\partial \underline{v}_1} + \frac{\partial \phi}{\partial \underline{r}_2} \cdot \frac{\partial F_3}{\partial \underline{v}_2} \right]. \tag{4.37}$$

In RHS of (4.36), ϕ is the potential between particles 1 and 2,
so that it is $\phi\left(\left|\underline{r}_1 - \underline{r}_2\right|\right)$, while in (4.37), we have to understand
the spatial dependence of the particular ϕ by its position in the
equation. Thus, on LHS it is always $\phi\left(\left|\underline{r}_1 - \underline{r}_2\right|\right)$, but on RHS it
first appears as $\phi\left(\left|\underline{r}_1 - \underline{r}_3\right|\right)$, and in the second term as
$\phi\left(\left|\underline{r}_2 - \underline{r}_3\right|\right)$. This should cause no further difficulty and saves
using a somewhat clumsy notation.

We have now to define some characteristic quantities. Let

$$r_o \quad = \quad \text{range of potential}$$

$$\langle\phi\rangle \quad = \quad \text{mean value of } \phi$$

$$u \quad = \quad \text{mean thermal speed}$$

$$L \quad = \quad \text{macroscopic length scale}$$

$$\tau \quad = \quad \text{macroscopic timescale.}$$

We can get some idea of the relative importance of terms in (4.36)
and (4.37), using these characteristic quantities, by the following
dimensional analysis. We need to relate F_3 to F_2 and F_2 to F_1.
Thus,

$$F_1 = \frac{1}{V} \int dr_2 \; dv_2 \; F_2.$$

Then, successive terms in (4.36) are in the ratio

$$1 : \frac{u\tau}{L} : nr_o^3 \; \frac{\langle\phi\rangle}{mu^2} \; \frac{u\tau}{r_o} \quad .$$

In (4.37), we can separate out, on LHS, the centre of mass motion
from the relative motion of particles 1 and 2.   We then obtain

$$1 \; : \; \frac{u\tau}{r_o} \; : \; \frac{u\tau}{L} \; : \; \left(\frac{<\phi>}{mu^2}\right) \left(\frac{u\tau}{r_o}\right) \; : \; (nr_o^3) \left(\frac{<\phi>}{mu^2}\right) \left(\frac{u\tau}{r_o}\right) ,$$

where the two terms involving ϕ on LHS are given equal weighting.

We see from this qualitative dimensional analysis that the RHS
can be treated as a perturbation in three distinct limiting situations, namely

$$nr_o^3 \; << \; 1 \; ; \; \left(\frac{<\phi>}{mu^2}\right) << 1 \; ; \; \frac{u\tau}{r_o} << 1 .$$

These obviously correspond to the three cases discussed earlier,
but with somewhat more precise definition.   Thus $nr_o^3 << 1$ is the
case of a short range interaction in a sufficiently dilute system.
The condition $\frac{<\phi>}{mu^2} << 1$ describes the situation where the mean
potential energy is much less than the mean kinetic energy and
hence has some virtue in being applied to a plasma, except that it
does not sufficiently emphasise the long-range nature of the
Coulomb interaction, as described by $u\tau/r_o << 1$.

It is useful here to introduce correlation functions for two
particles and for three particles, by the following definitions,
with obvious notation:

$$F_2 \, (1, \, 2) \; = \; F_1 \, (1) \, F_1 \, (2) \, + \, g(1, \, 2)$$

$$F_3 \, (1, \, 2, \, 3) \; = \; F_1(1) \, F_1(2) \, F_1(3) \, + \, F_1(1) \, g(2, \, 3) \, + \, F_1(2) \, g(1, \, 3)$$

$$+ \, F_1(3) \, g(1, \, 2) \, + \, h(1, \, 2, \, 3).$$

The object is to obtain equations involving F_1, g and h, rather
than F_1, F_2, F_3.   (For standardization of notation we shall now
write f for F_1.)

We then obtain, instead of (4.36) and (4.37)

$$\frac{\partial f}{\partial t} + \underline{v}_1 \cdot \frac{\partial f}{\partial \underline{r}_1} \; = \; \frac{n}{m} \int d\underline{r}_2 \, d\underline{v}_2 \, \frac{\partial \phi}{\partial \underline{r}_1} \cdot \left[\frac{\partial f(1)}{\partial \underline{v}_1} f(2) + \frac{\partial g}{\partial \underline{v}_1} \right] \qquad (4.38)$$

$$\left[\frac{\partial}{\partial t} + \underline{v}_1 \cdot \frac{\partial}{\partial \underline{r}_1} + \underline{v}_2 : \frac{\partial}{\partial \underline{r}_2} - \frac{1}{m} \frac{\partial \phi}{\partial \underline{r}_1} \cdot \frac{\partial}{\partial \underline{v}_1} - \frac{1}{m} \frac{\partial \phi}{\partial \underline{r}_2} \cdot \frac{\partial}{\partial \underline{v}_2} \right] g$$

$$= \frac{1}{m} \frac{\partial \phi}{\partial \underline{r}_1} \cdot \frac{\partial f(1)}{\partial \underline{v}_1} f(2) + \frac{1}{m} \frac{\partial \phi}{\partial \underline{r}_2} \cdot \frac{\partial f(2)}{\partial \underline{v}_2} f(1) +$$

$$+ \frac{n}{m} \int d\underline{r}_3 \, d\underline{v}_3 \left\{ \frac{\partial \phi}{\partial \underline{r}_1} \cdot \left[\frac{\partial f(1)}{\partial \underline{v}_1} g(2,3) + \frac{\partial g(1,2)}{\partial \underline{v}_1} f(3) + \frac{\partial h(1,2,3)}{\partial \underline{v}_1} \right] + \right.$$

$$+ \frac{\partial \phi}{\partial \underline{r}_2} \cdot \left[\frac{\partial f(2)}{\partial \underline{v}_2} g(1,3) + \frac{\partial g(1,2)}{\partial \underline{v}_2} f(3) + \frac{\partial h(1,2,3)}{\partial \underline{v}_2} \right] \right\} . \qquad (4.39)$$

To make further progress in the particular case of weak coupling, we have to relate the expansion parameter $\varepsilon = \frac{<\phi>}{mu^2}$ to the character-istic times τ_o and τ_1, the duration of a collision and the mean free time between collisions respectively. In terms of the other characteristic quantities defined earlier in this section, $\tau_o = r_o/u$ and $\tau_1 = \tau_o/\varepsilon^2$. The latter can be derived thus,

$$\tau_1 \simeq \frac{1}{n\sigma u} , \qquad \sigma = \text{ total scattering cross section}$$

$$\simeq \left(\frac{<\phi>}{mu^2} \right)^2 r_o^2 = \varepsilon^2 r_o^2 .$$

Hence, $\quad \tau_1 = \dfrac{1}{nu\varepsilon^2 r_o^2} = \left(\dfrac{1}{nr_o^3} \right) \cdot \dfrac{r_o}{u} \cdot \dfrac{1}{\varepsilon^2} \simeq \dfrac{\tau_o}{\varepsilon^2}$

provided that $nr_o^3 \simeq 0(1)$, that is we are not operating also in the short range, low density limit.

If we simply expand f and g as power series in ε,

$$f = f^{(0)} + \varepsilon f^{(1)} + \dots \dots$$

$$g = g^{(0)} + \varepsilon g^{(1)} + \dots \dots$$

and solve the equations for f and g, secular (i.e. unbounded) terms will result. If $t \simeq \tau_o$, we shall still get a satisfactory solution. However the whole point of a kinetic equation is to describe how a system evolves to equilibrium, a time requirement far in excess of τ_o. We then find that terms linear in t (or worse) dominate the series solution and lead to incorrect behaviour as $t \to \infty$.

We are led to adopt a different procedure, an expansion similar to that adopted in non-linear mechanics and developed by Bogoliubov and Krylov.[23]   This method allows a more general time variation, but requires an additional constraint to determine the additional functional dependence. This constraint allows us more flexibility in the solution. In the present context it is defined to be such that secular behaviour is eliminated at each successive stage in the expansion.

We therefore expand f and g as follows.

$$f = f^{(0)} (t_o, t_1, \dots .) + \varepsilon f^{(1)} (t_o, t_1, \dots .) + \dots .$$

$$g = g^{(0)} (t_o, t_1, \dots .) + \varepsilon g^{(1)} (t_o, t_1, \dots .) + \dots . \qquad (4.40)$$

The new time variables t_o, t_1, $\dots .$ are to be considered as independent variables, but finally are defined by

$$t_o = t, \quad t_1 = \varepsilon t, \quad t_2 = \varepsilon^2 t, \quad \ldots,$$

so that
$$\frac{dt_n}{dt} = \varepsilon^n .$$

We now proceed to expand equations (4.38) and (4.39).

To lowest order,

$$\frac{\partial f^{(0)}}{\partial t_o} = 0 \qquad f^{(0)} \text{ does not vary on } t_o \text{ scale}$$

$$\frac{\partial g^{(0)}}{\partial t_o} + (\underline{v}_1 - \underline{v}_2) \cdot \frac{\partial g^{(0)}}{\partial \underline{x}} = 0, \qquad (4.41)$$

where $\underline{x} = \underline{r}_1 - \underline{r}_2 =$ relative position vector of particles 1 and 2, and the motion of the centre of mass has been neglected.

However, if we assume $g = 0$ initially (to all orders in ε), we get $g^{(0)} = 0$.

First order

$$\frac{\partial f^{(1)}}{\partial t_o} + \frac{\partial f^{(0)}}{\partial t_1} = \frac{n}{m} \int d\underline{r}_2 \, d\underline{v}_2 \, \frac{\partial \phi}{\partial \underline{r}_1} \cdot \frac{\partial f^{(0)}}{\partial \underline{v}_1} f^{(0)} (\underline{v}_2) . \qquad (4.42)$$

Since $\phi = \phi\left(|\underline{r}_1 - \underline{r}_2|\right)$, and assuming $\phi \to 0$ as $|\underline{r}_1 - \underline{r}_2| \to \infty$, RHS vanishes. Equation for $f^{(1)}$ is therefore trivial, giving as solution

$$f^{(1)} (t_o) = f^{(1)} (0) - t_o \frac{\partial f^{(0)}}{\partial t_1} .$$

Note here the first appearance of secular behaviour as $t_o \to \infty$. We apply the constraint by choosing $\frac{\partial f^{(0)}}{\partial t_1} = 0$, giving therefore $\frac{\partial f^{(1)}}{\partial t_o} = 0$.

Second order

$$\frac{\partial f^{(2)}}{\partial t_o} + \frac{\partial f^{(1)}}{\partial t_1} + \frac{\partial f^{(0)}}{\partial t_2} = \frac{n}{m} \int d\underline{r}_2 \, d\underline{v}_2 \, \frac{\partial \phi}{\partial \underline{r}_1} \cdot \left[\frac{\partial f^{(1)}}{\partial \underline{v}_1} (1) f^{(0)} (2) + \right.$$

$$\left. + \frac{\partial f^{(0)}}{\partial \underline{v}_1} (1) f^{(1)} (2) + \frac{\partial g^{(1)}}{\partial \underline{v}_1} (1,2) \right]$$

$$(4.43)$$

$$\frac{\partial g^{(1)}}{\partial t_o} + (\underline{v}_1 - \underline{v}_2) \cdot \frac{\partial g^{(1)}}{\partial \underline{x}} = \frac{1}{m} \frac{\partial \phi}{\partial \underline{x}} \cdot \left(\frac{\partial}{\partial \underline{v}_1} - \frac{\partial}{\partial \underline{v}_2} \right) f^{(0)} (1) f^{(0)} (2).$$

$$(4.44)$$

To solve for $g^{(1)}$, we take the Fourier transform of $\underline{x}$ and the

Laplace transform of t_o, obtaining

$$\bar{g}^{(1)}(\underline{k}, \underline{v}_1, \underline{v}_2, p) = \frac{1}{m} \frac{\bar{\phi}(k) \, i\underline{k} \cdot \left(\frac{\partial}{\partial \underline{v}_1} - \frac{\partial}{\partial \underline{v}_2}\right) f^{(0)}(1) \, f^{(0)}(2)}{p \left[p + i\underline{k} \cdot (\underline{v}_1 - \underline{v}_2)\right]} .$$

$$(4.45)$$

Initially, $g^{(1)} = 0$.

The equation for $f^{(2)}$ can be similarly solved:

$$\bar{f}^{(2)}(p) = \frac{f^{(2)}(t_o = 0)}{p} - \frac{1}{p^2}\left[\frac{\partial f^{(1)}}{\partial t_1} + \frac{\partial f^{(0)}}{\partial t_2}\right] +$$

$$+ (2\pi)^3 \frac{n}{m} \int d\underline{k} \, \bar{\phi}(k) \, i\underline{k} \cdot \frac{\partial}{\partial \underline{v}_1} \int dv_2 \, \frac{1}{p} \bar{g}^{(1)}(-\underline{k}, \underline{v}_1, \underline{v}_2, p).$$

We must investigate the behaviour of $f^{(2)}$ as $t_o \to \infty$. The double pole at $p = 0$ leads to secular behaviour, since the Laplace transform of t_o is $\frac{1}{p}^2$. We must use freedom in $f^{(1)}$ and $f^{(0)}$ to eliminate such terms. To this end, we use the result

$$\lim_{p\to 0} p \, \bar{f}(p) = \lim_{t\to\infty} f(t)$$

to obtain asymptotic behaviour. We then require

$$\frac{\partial f^{(1)}}{\partial t_1} + \frac{\partial f^{(0)}}{\partial t_2}$$

$$= (2\pi)^3 \frac{n}{m} \int d\underline{k} \, \bar{\phi}(k) \, i\underline{k} \cdot \frac{\partial}{\partial \underline{v}_1} \int dv_2 \, \lim_{p\to 0} p\bar{g}^{(1)}(-\underline{k}, \underline{v}_1, \underline{v}_2, p).$$

$$(4.47)$$

Using the result $\quad \lim_{\eta\to 0} \frac{1}{x \pm i\eta} = P \frac{1}{x} \mp i\pi\delta(x)$

and retaining only the real part on RHS of (4.47)

$$\frac{\partial f^{(1)}}{\partial t_1} + \frac{\partial f^{(0)}}{\partial t_2} = 8\pi^4 \frac{n}{m} \int d\underline{k} \, |\phi(k)|^2 \, \underline{k} \cdot \frac{\partial}{\partial \underline{v}_1} \int d\underline{v}_2 \times$$

$$\left\{\underline{k} \cdot \left[\frac{\partial f^{(0)}}{\partial \underline{v}_1}(1) \, f^{(0)}(2) - \frac{\partial f^{(0)}}{\partial \underline{v}_2}(2) \, f^{(0)}(1)\right] \delta\left(\underline{k} \cdot (\underline{v}_1 - \underline{v}_2)\right)\right\} .$$

$$(4.48)$$

Equation (4.48) can be written in the form

$$\frac{\partial f^{(1)}}{\partial t_1} + \frac{\partial f^{(0)}}{\partial t_2} = C\left[f^{(0)}(1), f^{(0)}(2)\right], \qquad (4.49)$$

where C is short for the RHS of (4.48), but shows the structure of an operator acting on the zero order single-particle distribution functions $f^{(0)}(1)$ and $f^{(0)}(2)$.

Since $f^{(0)}$ is independent of t_1, (4.49) leads to secular behaviour, unless

$$\frac{\partial f^{(0)}}{\partial t_2} = C\left[f^{(0)}(1), f^{(0)}(2)\right]. \qquad (4.50)$$

The first result is again trivial, that we must have $\frac{\partial f^{(1)}}{\partial t_1} = 0$.
It is at this point that the method of Frieman and Sandri becomes productive, for as a by-product we have obtained a non-trivial equation for the t_2 dependence of $f^{(0)}$, namely (4.50). This equation is of the form of the general Fokker-Planck equation, an irreversible kinetic equation for $f^{(0)}(t_2)$. In this approach, then, the relevant kinetic equation is essentially a consistency condition for a well-behaved expansion without secular terms. There are several comments:

(i) In principle, we should be able to find higher order corrections to the Fokker-Planck equation, though there are considerable analytical difficulties.

(ii) It may be asked what would result instead of the Fokker-Planck equation, if we invoked the short-range or long-range inter-action criteria rather than weak coupling. As might be expected, the condition $nr_o^3 \ll 1$ leads to the Boltzmann equation. The case $u\tau/r_o \ll 1$ has been discussed extensively; the kinetic equation which results is known in the literature as the Balescu-Lenard-Guernsey (BGL) equation, which was first derived independently by Balescu[24], Lenard[25] and Guernsey[26]. It is hoped that in plasma physics the BGL equation will play the role of the Boltzmann equation in neutral gases, and that it will give an adequate description of relax-ation of a plasma to equilibrium. In its most general form the BGL equation is precisely equation (4.50) with one important modification, the replacement of $\phi(k)$ by $\phi(k)/\varepsilon(-\underline{k}, i\underline{k}\cdot\underline{v}_1)$. Note firstly that this implies that $\varepsilon(\underline{k}, \omega)$ never vanishes for any real $\underline{k}$, ω, so that the plasma must not be (linearly) unstable. Secondly, the effect of ε can be given a simple physical interpretation. If ϕ is the potential describing the interaction between a pair of particles in the absence of the plasma, and if the presence of the plasma can be described by a dielectric coefficient ε, then the potential is modified by just the factor $1/\varepsilon$. A more rigorous derivation is required

to supply the details, including a proper treatment of the Fourier-
Laplace transform $\varepsilon(\underline{k},\ p)$ of ε, but the final result amounts to the
simple description above, showing that the BGL equation is essen-
tially of Fokker-Planck form.   Because of the complexities of ε,
much work is still required to make a full assessment of the
properties of the BGL equation.

Chapter 5

MAGNETIC CONFINEMENT OF PLASMA

5-1 GENERAL PRINCIPLES

Magnetic confinement of plasma has for many years been considered
the most attractive method of confining plasma in laboratory thermonuclear fusion experiments, since very high magnetic pressures are
readily achieved.   A field of 0.5 tesla is sufficient to produce
a pressure of one atmosphere, and fields of up to 10T are realisable in practice.   We must remember, though, that the net pressure
exerted by the magnetic field on the plasma may be very much less
than $B^2/2\mu_o$, in the case of interpenetrating field and plasma, and
is only realised when plasma and field are completely separate,
which is usually undesirable for other reasons.   An important
parameter β is the ratio of plasma pressure to magnetic pressure,

$$\beta \; = \; \frac{2\mu_o\,p}{B^2} \; ,$$

which, in present-day experiments, is typically 25% in the 'high β'
experiments, 1% in Tokamak experiments and several orders of
magnitude lower in others.

The usefulness of a particular configuration can be evaluated in
various ways.   Theoretically, these can be separately discussed,
but experimentally there is no clear distinction.

   (i) The starting point is that it must be possible to set up an
        equilibrium state of the plasma.

  (ii) The stability of this equilibrium against gross magnetofluid
        perturbations must be investigated.

 (iii) Even if the configuration is macroscopically stable, residual
        microscopic instabilities, essentially involving the
        velocity distributions of the plasma ions and electrons,
        must remain at a sufficiently low level.

  (iv) One consequence of such microinstabilities is an enhancement
        of transport processes, leading to anomalous thermal and
        magnetic field diffusion, and ultimately loss of equilibrium
        A simple description was given by Bohm.[27] Microinstabilities

grow until their amplitude is limited by non-linear
processes, resulting in a steady level of turbulence, and a
cross-field diffusion coefficient is found to be

$$D_{Bohm} = \frac{1}{16} a_e^2 \Omega_e = \frac{\kappa T}{16eB} \, ,$$

the Bohm diffusion coefficient. Here, a_e is the gyromagnetic radius for an electron with thermal velocity and
Ω_e is the electron gyrofrequency so that $\Omega_e = \frac{eB}{m}$ and
$a_e = (\kappa T/m)^{\frac{1}{2}}/\Omega_e$. Notice that the Bohm diffusion coefficient
is inversely proportional to B. A simpler picture in which
diffusion of electrons across the magnetic field is due to
electron-ion binary collisions leads to a 'random walk'
classical diffusion coefficient

$$D_{classical} = a_e^2 \nu_{ei} \, ,$$

where ν_{ei} is the electron-ion collision frequency. Since
the latter is (essentially) independent of B, this classical
coefficient is proportional to $\frac{1}{B^2}$.
This brings out one important aspect of magnetic confinement
experiments, the possible existence of scaling laws.
Suppose that we have obtained results from an experiment and
we wish to consider designing a new device on the basis of
these results. Clearly we shall be greatly helped if we
can predict the effect of changing the physical and
geometrical parameters, all other things being equal. Since
ultimately we desire as long a containment time as possible,
it is of crucial importance to know the rate of diffusion in
the new system. Bohm diffusion predicts one scaling law as
the magnetic field is increased, classical diffusion predicts
a more favourable law. One objective of present-day
experiments is to decide which, if either, is correct, and
under what conditions.

(v)   Other sources of anomalous transport coefficients also occur,
      such as the effect of finite toroidal geometry.

The remaining sections deal with specific examples of magnetic
confinement systems, namely, <u>open systems</u> in which magnetic field
lines eventually leave the confined plasma, and <u>closed systems</u> in
which the field lines remain in the plasma.

5-2 OPEN SYSTEMS

(i) Cylindrical axial pinch

The earliest magnetic confinement system, the 'cylindrical pinch'
is based on a simple theoretical analysis given by Bennett.[28]
The configuration was described briefly in §3-4.   Our starting
point is equation (3.12), for the case of an axial current density
$j_z(r)$ generating the confinement field $B_\theta(r)$ so that

$$\frac{dp}{dr} = -\frac{B_\theta}{\mu_o r}\frac{d}{dr}(rB_\theta). \tag{5.1}$$

Let the radius of the plasma column be r_o.   Multiply (5.1) by r^2
and integrate over r from 0 to r_o:

$$\int_0^{r_o} r^2\frac{dp}{dr}\,dr = -\frac{1}{\mu_o}\int_0^{r_o}(rB_\theta)\frac{d}{dr}(rB_\theta).$$

Hence,

$$\left[r^2 p\right]_0^{r_o} - 2\int_0^{r_o} rp\,dr = -\frac{1}{2\mu_o}(rB_\theta)^2\bigg|_{r=r_o}. \tag{5.2}$$

We suppose $p = 0$ at $r = r_o$ and that ions and electrons are at
uniform temperature T.   Then

$$p = 2n\kappa T$$

with $n(r)$ =   density of electrons

              =   density of ions.

Hence

$$2\kappa T\int_0^{r_o} 2r\,n\,dr = \frac{1}{2\mu_o}(rB_\theta)^2\bigg|_{r=r_o}. \tag{5.3}$$

Let N = number of electrons per unit length of cylinder

$$= 2\pi\int_0^{r_o} n\,r\,dr.$$

Hence LHS of (5.3)   = $2\kappa TN/\pi$.

Using Maxwell's equations,

$$\frac{1}{r}\frac{d}{dr}(rB_\theta) = \mu_o j_z.$$

Hence

$$(rB_\theta)\bigg|_{r=r_o} = \mu_o\int_0^{r_o} j_z\,rdr = \frac{\mu_o I}{2\pi},$$

where I = total axial current.

Substituting these results into (5.3) gives

$$I^2 = \frac{16\pi}{\mu_o} \kappa TN,$$ (5.4)

which is the famous Bennett relation. It shows that, provided
the pinch is stable and possesses cylindrical symmetry, the plasma
temperature is proportional to the square of the discharge current
and inversely proportional to the particle line density. This
again has the feature of a scaling law. It does not depend on
the particular radial distribution of density or current.

(ii) Theta pinch

The theta pinch configuration has also been briefly described in
§3-4. Devices based on this configuration are generally known as
Thetatrons. It is of interest to describe in more detail the
sequence of operations usually followed in an experiment. A low-
inductance, high-energy capacitor bank is discharged through the
single turn metal strap which encircles the cylinder, thus creating
a rapidly rising axial magnetic field. This induces an azimuthal
skin current j_z in the plasma which excludes the magnetic field
from the plasma, thus developing the maximum pressure, and rapidly
radially compressing the plasma. The coil current can be arranged
to rise so rapidly that a shock wave advances radially inwards
ahead of the magnetic 'piston'. This results in shock heating of
the plasma, which is further heated by the subsequent slower phase
of adiabatic compression. At present a Thetatron may typically
have a rise time of order 10μs, resulting in a density $n \simeq 10^{22}$ m^{-3}
and a temperature $T \simeq 10^7$ K. The containment time is limited by
the leakage of plasma out of the ends of the device and may reach
several μs.

It is not the intention here to embark on a long review of the
merits and faults of Thetatrons, but a few words are in order.
End loss is finally the most serious drawback and precludes
adoption of a Thetatron as a fusion reactor. However in early
observations lack of cylindrical symmetry led to drifting of the
plasma to the wall and filamentary breakup of the plasma. The
first fault was cured by improved coil design and the latter, due
to rapid axial rotation of the plasma, which in turn was the
result of a strong radial electric field, was suppressed by careful
programming of the rate of compression in the initial phase, as
well as by lengthening the Thetatron. In the absence of these

faults, classical diffusion across the field has been observed.

(iii) Magnetic mirror trap

The principles of an axisymmetric magnetic mirror trap were given
in §2-4. This type of system suffers immediately from end loss
problems, but it has other built-in ailments. The magnetic field
intensity in a direction normal to the field lines decreases in
going radially outwards from the axis, in the central region of the
trap where the plasma is localised. This transverse magnetic
field gradient drives an instability which leads to an enhanced
loss of plasma across the field far in excess of the classical
rate. The mechanism for these instabilities has been discussed in
§2-6-1; the perturbations form a fluting of the plasma surface,
caused by local interchanging of plasma and magnetic field. A
reversal of the magnetic field gradient suppresses the interchange
instability. This can be achieved by the addition of a quadrupole
magnetic field, of such a strength that a magnetic well is
established. This has the property that $B = |\underline{B}|$ increases in all
directions from the centre of the well, and can be realised
experimentally by introducing four straight conductors parallel to
the axis of the magnetic trap and symmetrically disposed (see
Fig. 5.1). Currents flow in alternate directions in these
conductors. A striking demonstration of the stabilisation was
given by Ioffe.[29] The trap was initially operated with zero
current in the bars, and the containment time was found to be of
order of several μs. No significant improvement occurred when
current was switched on in the bars until a certain critical
current I_c was reached. At this point the plasma became much more
quiescent and the containment time was dramatically increased by
approximately three orders of magnitude. The value of I_c
coincided with the formation of a magnetic well.

From the discussion in §2-4, we see that the velocity distribution of particles contained in a magnetic mirror trap cannot be
Maxwellian, since particles in the ' loss cone' are immediately
lost. As well as providing a leak, by diffusion into the loss
cone, the anisotropy in the distribution function can also result
in a microinstability, akin to a two-stream instability (cf. §4-2-2),
and known as the loss-cone instability.

As this short discussion shows, the apparent simplicity of the
axisymmetric magnetic mirror trap suffers from many fundamental

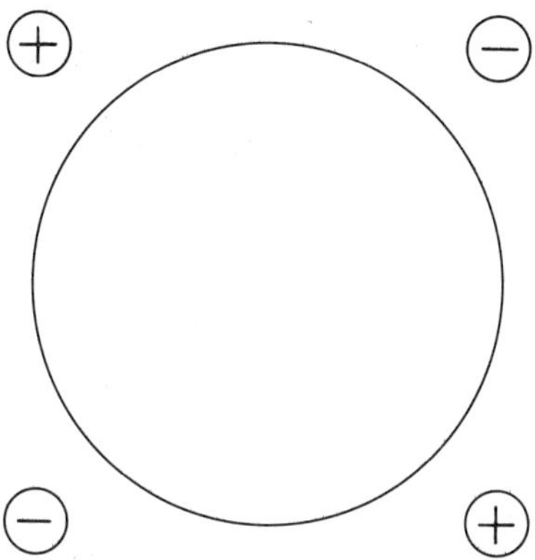

Fig. 5.1 Cross section of mirror trap with stabilising
quadrupole field

difficulties, so that although it has not completely gone out of
favour as a potential fusion device it is unlikely to be a main
contender.

5-3 CLOSED SYSTEMS

The basic idea of toroidal equilibria was outlined in §3-4.   In
this section a short review will be given of three main classes of
toroidal systems, (i) high β toroidal experiments, (ii) Tokamaks,
(iii) the Levitron concept.   Their common features are the absence
of end effects and the existence of magnetic surfaces, but they
also differ in many essential respects, which are outlined below.

(i) High β toroidal experiments

The forerunner of this important class of containment systems was
ZETA, which began operation at AERE Harwell in 1957.   The basic
idea was simple enough.   According to the Bennett relation, which
in a slightly modified form is applicable to toroidal pinches, the
higher the total current the higher the temperature, $T \propto I^2$.
Initially, containment in toroidal pinches was transitory, in which
the magnetic field was produced entirely by a toroidal current.
The breakdown occurred due to a magnetofluid instability described
as the 'kink' instability.   The self-field due to the current
consisted entirely of a poloidal component, and any spontaneous
kinking or buckling of the plasma column was accompanied by an
inevitable strengthening of the field on the inner side of the
kinking, leading to further enhancement.   These instabilities were
fully discussed in §3-6, the 'kink' mode corresponding to m = 1.

Another dangerous perturbation, which corresponds to the axi-symmetric mode m = 0, occurs where the plasma column narrows. The field, which has approximately a $\frac{1}{r}$ dependence on radius, again strengthens, further constricting the plasma until a disconnection occurs.

This simple description is strictly applicable only to cylindrical pinches, but with care can be applied also to toroidal systems. The instabilities are largely suppressed by introducing initially a toroidal field, which is trapped in the plasma when the main discharge occurs. It is the combination of a toroidal and poloidal component which generates magnetic surfaces (cf. §3-4).

Having now achieved a stabilised toroidal configuration, satisfying the Bennett relation, we can now see immediately that any attempt to increase the current to very high values will eventually fail. We suppose that initially there is only a toroidal magnetic field B_ϕ. Introduction of a small toroidal current produces a poloidal field $B_\theta \ll B_\phi$. Let R be the major radius of the torus. Then the field lines satisfy the differential relation

$$\frac{R\,d\phi}{B_\phi} = \frac{r\,d\theta}{B_\theta} .$$ (5.5)

Hence, in a single circuit around the torus, $d\theta$ increased by $2\pi \frac{R\,B_\theta}{rB_\phi} = \frac{2\pi}{q}$. q is sometimes referred to as the 'safety factor'. If q = 1, the field line closes on itself. If q > 1, a field line instead generates a magnetic surface. Hence, increasing the current and thereby decreasing q from, initially, a value much greater than unity, eventually leads to the configuration q = 1, which rapidly disintegrates. This condition was first pointed out by Kruskal and Shafranov in the 1950's, but its full signif-icance was not realised until the Tokamak experiments more than a decade later.

The requirement q > 1 is strictly applicable only for low β. In the high β toroidal experiments we in fact have the opposite situation, with $B_\theta \simeq |B_\phi|$ and q < 1.

The present interest in high β experiments stems principally from the realisation that ultimately the efficiency of a fusion reactor depends critically on having a high enough β, certainly in excess of 10%. The reason is simple. It costs energy to maintain large containing magnetic fields, and so the higher the β the more

plasma is contained for a given field. This in turn pays back
proportionately more thermonuclear energy. Of course, this simple
argument assumes similar containment times in the various systems
under consideration.

One high β device currently of great interest is the Reversed
Field Pinch (RFP) at Culham Laboratory. It earns its name from
the fact that the toroidal magnetic field has a reversal of sign
in the outer region of the discharge, which is beneficial to
stability (Suydam's stability criterion is then more easily
satisfied).

The magnetic fields are carefully programmed and theoretically
the plasma should be formed in a stable configuration. Localised
Suydam modes are not observed, while the resistive tearing mode
(cf. §3-7) is identified experimentally and suppressed by a
reduction in the compression ratio in accord with theory.

Detailed studies both in the high β experiment (without field
programming) and using the original ZETA results, show that the
plasma by itself generates a RFP configuration spontaneously, in
substantial agreement with theoretical predictions by Taylor.[30]
The natural generation of a reversed field is almost always
observed in pinch experiments with longitudinal flux-conserving
shells when the ratio $\Theta = B_\theta$ (wall)/B_ϕ (average) exceeds 1.4.
Qualitatively, self-reversal occurs because poloidal currents are
generated which increase B_ϕ towards the axis and, to conserve flux
within the shell, the outer field falls and can reverse. Taylor
shows that, given small, unspecified, dissipation, a pinch discharge
will always tend to relax to a force-free ($\beta = 0$) minimum energy
configuration.

The distribution is characterised by Θ which determines the
ratio $\Phi = B_\phi$(wall)/B_ϕ (average). When $1.2 < \Theta < 1.56$, $\Phi < 0$ and
the configuration is the axisymmetric RFP. Furthermore, $\Theta = 1.56$
is an upper limit for $\beta = 0$; and beyond, helical deformations
instead are generated. While this theory is strictly applicable
only to $\beta = 0$ states, it is observed that high-β states which are
stable according to ideal magnetofluid theory lie close to the
$\beta = 0$ states predicted by Taylor. Thus a plasma initially unstable
will by itself move almost to the desired stable high-β state.
(ii) Tokamaks
Tokamak devices are toroidal plasma systems characterised by high
total thermal energy in the plasma and a long energy containment

time. Currently, more ambitious Russian and American Tokamaks
are coming into operation or are still on the drawing-board,
together with a possible European project, the Joint European
Torus (JET). However, perhaps the most notable Tokamak device was
the system T3 at the Kurchatov Institute in Moscow, a torus with
major radius 1m, and minor radius 0.175m. The main toroidal
field B_ϕ could be increased to 3.7 tesla, and the current pulse I
of up to 180 kA had a duration of 70ms. An important feature was
the application of a vertical magnetic field, to control plasma
displacement off axis.

The main parameters of the plasma generated in T3 are: density
$n = 10^{19} m^{-3}$, electron temperature $T \simeq 1 - 2 \times 10^7$ K, containment
time $\tau \simeq 20$ ms. Experiments showed the following encouraging
properties of τ:

   (i)   It does not depend on the strength of B_ϕ.

  (ii)   It increases with n.

 (iii)   It increases with the total current I.

  (iv)   It is not reduced with increasing T.

   (v)   At high T, τ differs significantly, being larger by a factor
      $\simeq 50$, from the characteristic value for the containment time
      determined by the Bohm formula, obeyed by most previous
      systems. This is expected to be connected with the
      quiescent operation of Tokamaks, and hence a classical-like
      behaviour, in contrast with the high degree of turbulence
      experienced by most other systems. It must be noted, though,
      that the diffusion observed (or inferred) in T3 is by no
      means purely classical, but approaches a factor 5 over
      classical though obeying the <u>classical</u> scaling law $\propto \frac{1}{B^2}$.

At first glance, Tokamak devices seem very simple. In practice,
the Tokamak plasma lives a complex life. For example, the
principal mechanism responsible for energy loss in Tokamaks is
high electron thermal conductivity. Its value is up to two orders
of magnitude greater than that predicted by Galeev and Sagdeev
(cf. §2-4). The mechanism of this anomalous transport of heat is
unclear, and much theoretical and experimental work is still
required.

At the same time, large new installations are being planned, so
that much effort is also required in studying (i) geometrical
effects, such as finite aspect ratio (the ratio of major to minor
radius of the torus), and non-circular cross sections; (ii) new

heating techniques, such as radio-frequency heating, fast-neutral
injection, and others.

In conclusion, the Tokamak approach appears to offer the best
hope for a successful outcome to the quest for a magnetic containment system for a fusion reactor. As a measure of its success so
far, the ion temperatures in the next generation of Tokamaks should
be well in excess of 10^7 K, while the $n\tau$ value (cf. §1-1) should be
between 10^{18} and 10^{19} m^{-3} s, which is within striking distance of
the Lawson criterion for a deuterium tritium mixture.

(iii) The Levitron concept

We now return briefly to the 'hardcore' concept, mentioned
in §3-6-3 (page 61), which on the basis of ideal magnetofluid
theory offers the possibility of absolute stability. Early
experiments involving cylindrical devices showed some degree of
stability as compared with pinched discharges, but end effects and
finite resistivity masked any real advantage. Early toroidal
systems fared no better, and showed conclusively that the onset of
resistive instabilities was very rapid. One weakness of these
early experiments in toroidal geometry was that physical supports
were required to hold the 'hardcore' ring in place, and it was
observed that these supports were a source of trouble. Magnetic
levitation was initially out of the question, but in recent years
improved computational techniques as well as the more widespread use
of superconducting materials has led to a revival of the concept of
magnetic levitation. One such system is the Culham Laboratory
LEVITRON. In this device, a magnetic field configuration of high
shear is generated in a toroidal vacuum region, partly by a current
of 1 MA along the major axis, to give a large toroidal component of
magnetic field, and partly by a current of 0.5 MA induced in an
inner toroidal ring producing a strong poloidal component, the
whole system of conductors being cooled to superconducting temperatures. A supplementary set of toroidal coils provide a vertical
field, for control of the ring and also, if desired, to produce a
magnetic well between the ring and the major axis. The whole
magnetic field, though very complex, possesses axial symmetry.

Plasma formation is a two-stage process; firstly, an externally
injected beam of 5 keV neutral atoms is ionised within the region
between ring and axis, partly by Lorentz ionisation in the strong
magnetic field and partly by collision with the neutral background

gas.   In the second stage, a general thermalisation takes place,
involving collisions between the fast ions and the background of
electrons and neutrals, until finally a plasma at a temperature
around 2×10^6 K is formed.

Theoretically, much work is proceeding in various aspects; for
example, the motion of a charged particle in such a magnetic field
can be very complicated, and the additional requirement of computing
self-consistent electric fields leads to a challenging problem.
The thermalisation stage can best be tackled by solving an appropriate Fokker-Planck equation, and the final stage in which a plasma
is in quasi-equilibrium offers many opportunities for enthusiastic
theorists.

Chapter 6

ASTROPHYSICAL PLASMAS[31,32]

6-1 INTRODUCTION

In some respects, this chapter is the odd man out, since the rest
of the book emphasizes the author's main interests in plasma physics,
centred on fusion research.   It is aimed at widening the reader's
horizons in plasma physics, and presents a brief, and obviously
incomplete, look at plasma phenomena which occur outside the confines
of the laboratory.

Although it may be claimed that the universe is almost entirely
in the plasma state, plasma effects by no means always dominate.
Three representative areas are presented where plasma physics does
have an important role to play in our understanding of astrophysical phenomena:

(1)   The solar wind and its interaction with the earth's magnetic
      field.
(2)   The magnetofluid nature of solar flares.
(3)   Radiation from pulsars.

Before proceeding with the programme outlined above, one point
that should be stressed is the wide range of physical conditions
which occurs in the cosmos.   For example, intergalactic plasma,
for which there is as yet no real evidence, is not likely to have
a density exceeding 10 m^{-3}.   In our own galaxy, interstellar
plasma has a density of order 10^6 m^{-3}, while stellar surface
densities may be as high as 10^{22} m^{-3}.   Again, a similar range
occurs in magnetic field intensities;   interstellar magnetic fields
are of order 10^{-10} T, with somewhat higher fields occurring in
radio galaxies.   The sun has relatively strong magnetic fields
localised in sunspots, $\simeq 0.1$ T, falling to 10^{-4} T as a typical
surface value, not unlike the earth's magnetic field.   At the top
end of the scale it is believed that pulsar atmospheres can have
magnetic fields as high as 10^8 T and above.

A few introductory remarks about shock waves. In a neutral gas of
pressure p, density ρ, temperature T and ratio of specific heats γ,
sound waves propagate with speed $C_S = (\gamma p/\rho)^{\frac{1}{2}} \simeq \rho^{\frac{1}{2}(\gamma-1)}$. Thus C_S
increases with density, and it was first observed by Stokes that
in an ideal fluid this leads to the development of a shock wave,
since a compression wave produces an increase in density. Hence,
behind the front the pressure wave is transmitted more rapidly
than on the front. The pressure gradient continually steepens
until it becomes a discontinuity, which is the shock front. This
conclusion has no physical meaning, as is well-known to students
of hydrodynamics. The discontinuity can be removed by taking
account of heat conduction and viscosity. So far, dissipative
processes have not been mentioned, but it is clear that dissipation
of energy increases as the wave front steepens. As a result, a
balance is reached between non-linear motion (the steepening of
the pulse) and energy dissipation, and instead of a discontinuity
a transition layer forms across which this balance is achieved.

In a neutral gas the transition layer has a thickness typically
of several mean free paths in the gas. Normally, this is so
small compared with any macroscopic length in the gas that, in
practice, it is indistinguishable from a discontinuity. We thus
still have a shock wave and the density, pressure, temperature
and flow velocity of the gas may be quite different on either side
of the shock. The dissipative effects determine the <u>structure</u> of
the shock layer, but not the state of the gas on either side of
the shock and outside the layer.

Relations between the two states can be obtained using the
conservation laws of mass, momentum and energy, and are known as
the Rankine-Hugoniot relations.

The same arguments can be applied to the case of a magnetofluid,
with the extra effect of finite electrical conductivity. However,
one essential difference in a high temperature plasma is that the
conventional collisional mean free path may be very long. In a
typical laboratory plasma experiment it may even exceed the
dimensions of the plasma itself. We should therefore not expect
to see a shock wave in such a situation. However, the presence
of a strong magnetic field may alter this conclusion, since the
gyromagnetic radius is a characteristic length for charged particle

motion normal to the field, and in certain circumstances may be
viewed as an effective mean free path for motion in that direction,
though not parallel to the field.

The magnetic field in the vicinity of the earth is approximately
a dipole field, which would extend outwards indefinitely if the
earth were in a vacuum.   However, it lies in the continuously
streaming solar "wind" of energetic charged particles, mainly
protons and electrons.   This wind has all the usual plasma
characteristics;   in particular, it has low electrical resistivity
and so compresses the earth's magnetic field $\underline{B}$ until the solar wind
pressure just balances $B^2/2\mu_o$.   The result is that the earth,
together with its magnetic field and radiation belts, is restricted
inside a cavity – the magnetosphere – outside which is the solar
environment.   Around 1960 it was realised that the magnetosphere,
of radius $\simeq 10^5$ km, would represent an obstacle to the solar wind.
Since the speed of the solar wind is 400 km s^{-1} and the sound speed
is 50 km s^{-1} the Mach number is 8, suggesting conditions for a
steady shock wave to exist.   Further, for such a high Mach number,
it would be expected that the shock should be detached from the
magnetosphere, by analogy from aerodynamic experiments.   It is then
possible to predict that a bow shock should occur at 15-20R (earth
radii), taking the radius of the magnetosphere to be 10-15R.

In the case of a shock wave produced by a supersonic object in
the earth's atmosphere, the phenomenon is dominated by interatomic
binary collisions, and the mean free path for such collisions is
typically a millimetre or less.   In contrast, for the bow shock in
the solar wind, the mean free path for Coulomb scattering between
protons, the equivalent mechanism, is approximately 10^9 km, of the
order of the solar distance, four orders of magnitude greater than
the magnetosphere radius.   It is fair to claim that binary
collisions between protons play a negligible role in the bow shock
phenomenon.

Nevertheless there is overwhelming experimental evidence for the
existence of a bow shock.   In 1964 the Explorer and IMP research
satellites included in their experimental programme measurement
studies of energetic electrons in the range 1-10 keV located outside
the magnetosphere.   A large flux of electrons is observed, with a
sharp reduction at about the expected position for the bow shock.

Magnetic field measurements show a sudden transition from the
dipole field within the magnetosphere to a very disturbed magnetic

field, followed by a second sharp transition and fall off of the
magnetic field intensity by a factor 2-3, as the satellite moves
radially outwards.   The final region is the undisturbed inter-
planetary medium.   The second transition occurs at about the same
location as the sudden change in electron flux.

The accumulated data may be briefly summarised, thus:

(i)   <u>Magnetospheric boundary</u>

Sudden appearance of plasma, mainly at 10R but flaring out
to 15R.

(ii)   <u>Transition region</u>

Wide energy spectrum, large fluctuations in flux and
direction of the plasma stream.   The region extends to
15-20R.

(iii)   <u>Shock wave</u>

The transition region (ii) gives way suddenly to one of
smoother flow, narrow energy spectrum and flux localised in
the solar direction.   The shock thickness is observed to be
in the range 10-100 km.

(iv)   <u>Solar wind</u>

Uniform plasma flow, with density $3 \times 10^6 \text{ m}^{-3}$.

Typical temperature measurements suggest $T \simeq 10^5$ K on the solar
side of the shock wave and 10^6 K on the earth side.

These observations are consistent with the existence of a shock
wave.   It seems fairly certain that the mechanism cannot be
collisional, that is, due to binary Coulomb collisions.   We
therefore have to look for collective processes.   Let us first
consider the characteristic lengths available.   In the absence of
a magnetic field there are only two simple length scales, the
collisional mean free path already mentioned, of order the solar
distance, and the screening length $\lambda \simeq 10$ m.   In between, consistent
with the observed magnetic field strengths, the electron and ion
gyromagnetic radii are in the range 1-100 km, which is the range of
interest for the thickness of the bow shock.

In order to put these results in perspective with measurements
in laboratory high temperature plasma, we note that, in the latter,
transport coefficients such as viscosity and electrical resistivity,
calculated from two-particle interactions, are many orders of
magnitude too low.   In recent years, experiments have concentrated
on propagation of small amplitude waves and the development of

120

instabilities. There is considerable evidence from these experiments that dissipation rates are much more rapid than would be
expected purely from random interparticle collisions. This is
still a feature which is not too well understood, but it is
expected that any explanation will be in terms of the interaction
of collective plasma modes, a topic of great importance in plasma
turbulence theory.

One of the most direct ways of studying dissipative mechanisms
is to investigate the generation of shock waves in high temperature
plasma, and the structure of these shocks. For, as we have already
inferred, if a shock _does_ occur, the conservation laws require the
existence of a dissipative mechanism (for example, entropy across
a shock must increase).

Several theoretical accounts of the bow shock have been given,
none entirely successful. Some invoke a non-linear phase of the
well-known two-stream instability. A turbulent spectrum of plasma
waves is set up which scatters incoming particles and enables the
incoming relatively cold supersonic plasma to be transformed into
hot subsonic plasma.

6-3 SOLAR FLARES

A solar flare is a very complex process. It is one of the most
dramatic events that regularly occur in the solar system, represent-
ing a catastrophic release of energy up to 10^{25} J per event. A
large variety of phenomena are associated with a flare:

(i)   _Radiative emission_

X-ray, ultraviolet, optical and radio-frequencies are all
rapidly enhanced.

(ii) _Energetic particles_

Charged particles are accelerated and emitted with energies
up to the relativistic range.

(iii) _Plasma dynamics_

Phenomena associated with flares include gross mass motion,
explosive instabilities, and shock waves, which may all be
categorised as magnetofluid phenomena.

To account for this wide range of physical effects, it is useful
to consider a flare to be made up of three components, though in
reality these will be closely inter-related:

(i)   _Optical flare_

A relatively cool plasma, with a temperature of order 10^{4} K,

density 10^{19} m^{-3}, somewhat denser than the surrounding
solar atmosphere.

(ii)   Hot, high energy flare

Plasma at a temperature between 10^7 K and 10^8 K, and density
10^{16} m^{-3}, emitting thermal X-rays.

(iii)   Energetic particle flare

This third component is responsible for non-thermal highly
energetic electrons and impulsive bursts of hard X-rays.

There are quite distinct phases in the development of a flare:

(i)   Pre-flare phase

This is marked by various types of activity, such as small
precursor filamentary flares, expansion of coronal arches,
and gradual bursts of microwave emission.

(ii)   Flash phase

The onset of the flare proper is accompanied by a rapid
increase of optical, microwave and X-ray emission, the latter
with energy in excess of 20 keV. This indicates that
considerable acceleration of charged particles is already
occurring.

(iii)   Decay phase

Various emissions subside, until pre-flare conditions are
re-established.

Phase (i) may last up to one hour, phase (ii) in which most of
the spectacular events occur lasts 5-10 minutes, and phase (iii)
varies greatly from half an hour up to several hours.

The suddenness of the onset of the flash phase suggests that the
main flare is an explosive process. In the pre-flare phase, a
gradual stressing and storing of energy occurs, until conditions
are right for a plasma instability to develop, deriving its energy
from the magnetic field and giving rise to Joule heating and
charged particle acceleration.

A theoretical account of the flare mechanism must therefore
explain the origin, storage and release of the primary flare energy.
The main requirements, and possible theoretical models, can be
summarised thus:

(i)   A magnetic field must play a central role.

(ii)   The pre-flare phase must possess more energy than the decay
phase. This surplus energy must become available for
heating, production of high energy particles, and large-scale

mass motion.

Since, for given boundary conditions on the magnetic field,
the plasma state with zero current density is the minimum
energy state, it follows that in the pre-flare phase finite
plasma currents flow. Further, the plasma has low β (as
defined for example in §3-6-1), that is, the magnetic field
energy is dominant, so the so-called force-free approximation
is valid. In such a case, the equilibrium equation

$$\nabla p = \underline{j} \times \underline{B}$$

becomes simply

$$\underline{j} \times \underline{B} = 0,$$

since ∇p is considered negligible. As a consequence, current
and field are parallel.

If then a force-free configuration with finite currents flowing goes to a state with less (or zero) current, energy
becomes available. One instability that can cause this
kind of change of state is the tearing-mode instability,
which was discussed in §3-7.

(iii)   The existence of the flash phase must be due to a rapid plasma
instability. The evidence is convincing, for not only does
it explain the suddenness of the onset of the flash phase,
but in some cases flares can be triggered. Further, the
instability must give rise to Joule heating and acceleration
of charged particles, and the energy released in the process
comes from the magnetic field.

The reader is referred to more extensive review of solar flares
for a more detailed account of observations and theoretical models.

6-4 PULSARS

The discovery in 1967 of the first pulsar has opened a fascinating
new chapter in astrophysics. We shall restrict our discussion to
associated plasma phenomena, and the physical models which try to
explain them. We shall not try to elaborate on the detailed
structure of the pulsar core, but will simply accept that it is a
neutron star, consisting mainly of a degenerate Fermi distribution
of neutrons at a density some 14 orders of magnitude greater than
normal densities. Radiation from pulsars, whether optical or at
radio-frequencies, is emitted in pulses, remarkable for their
accurate periodicity, and it is generally agreed that only stellar

rotation, combined with beamed emission, can give a satisfactory
explanation. The short periodicities observed, of order one
second or less, exclude as possible candidates all but neutron
stars, since only the latter can rotate sufficiently fast without
disrupting because of their relatively small geometrical size.

The pulsar atmosphere is currently the most fascinating aspect
for plasma physicists. For, although the periodicity of rotation
is typically one second, the pulsed radiation is observed throughout the spectrum, radio-frequencies, optical frequencies, and even
in the X-ray band. We must therefore look for a mechanism which
transforms the rotational energy of the neutron star into pulsed
electromagnetic radiation. One popular model is to assume the
existence of a magnetosphere outside the core in which coherent
radiation is generated and beamed. This is a problem of electrodynamics combined with relativistic plasma physics.

Since we cannot send a probe to a pulsar, we can make progress
in understanding how it works by a combination of inspired guesses
and good physical reasoning. Let us first focus our attention on
the magnetic field of the pulsar. Starting with a typical value of
stellar magnetic fields, observed for example in the sun, and taking
account of conservation of magnetic flux during the initial collapse
of the neutron star, we would expect a surface magnetic field in
excess of 10^8 T. Now, when a normal atom is placed in a magnetic
field of 10^5 T, the electron gyromagnetic radius is comparable to
the size of the atom. An increase in the field to 10^8 T must
therefore result in considerable distortion, with the atomic electrons
essentially confined to the surface of a cylinder with its axis
aligned to the magnetic field. Matter in bulk therefore possesses
highly anisotropic properties. An important consequence is to
inhibit the removal of charges from the surface of a neutron star,
since an estimate of the work function is 1 keV. Consequently, we
expect a pulsar plasma atmosphere to form, though not necessarily
with equal numbers of positive and negative charges. The low
electrical resistivity of the plasma atmosphere, which rotates with
the same angular velocity $\underline{\Omega}$ as the pulsar core, requires charge
separation to occur until the resulting electric field $\underline{E}$ and the
Lorentz force (per unit charge) $\underline{v} \times \underline{B}$ satisfy the condition

$$\underline{E} + \underline{v} \times \underline{B} = 0,$$

where $\underline{v} = \underline{\Omega} \times \underline{r}$, for position vector $\underline{r}$ in the atmosphere.

Ultimately, as we move radially outwards, we reach a position
such that v = c, the 'light cylinder' where co-rotating particles
would acquire the speed of light.   This requires a relativistic
treatment but it is clear that beyond the light cylinder, of radius
$\frac{c}{\Omega}$, charged particles and magnetic fields must be continually swept
away into interstellar space.

Charges leaving the pulsar atmosphere in the vicinity of either
polar region escape along field lines, and are all of the same sign.
To maintain equilibrium, there must be a corresponding flow of
charge into these regions.   The resulting model becomes very complex, involving for example flow of charge of one sign through an
oppositely charged region.

A model such as the one briefly outlined above can be put on a
more quantitative basis, and the rate of energy loss outside the
light cylinder can be calculated.   Now, observations show that
pulsar periods show a systematic increase, presumably as the
pulsar core rotation slows down.   For the Crab nebula, the rate
of loss of rotational energy from the central pulsar corresponds
closely with the energy required to maintain the observed synchro-
tron radiation from the entire nebula.

While this check is very satisfactory, the actual radiation
mechanism of pulsars has not been positively established, though
several models have been proposed.   Some mechanism for accelerating
bunches of charged particles is necessary.   One attractive model
uses the fact that a group of electrons, in which each particle
moves almost parallel to a curved magnetic field line, will radiate
coherently at wavelengths greater than the characteristic size of
the bunch.   For a relativistic group of electrons, the radiation is
emitted tangentially to the field line.   A succession of such
groups, moving outwards along the magnetic polar axis of the pulsar,
will radiate a beam of pulsed radiation, as observed, provided that
the rotation axis and magnetic axis are not aligned.

Much work still requires to be done before we can claim to
understand at all fully the complex physical phenomena occurring
in pulsar atmospheres.

Chapter 7

SUPERDENSE PLASMA

7-1 INTRODUCTION

During the last two or three years, intensive studies of the laser
compression of matter to ultra-high densities have been initiated,
primarily in the USSR and USA. The impetus for this research is
the possibility of a novel approach to thermonuclear fusion.

In the magnetic confinement approach to fusion, the density n is
limited to the range $10^{19} - 10^{22}$ m^{-3}, in part because of the limit
on generating large magnetic fields, but also because of the basic
properties of gaseous plasma at thermonuclear temperatures. The
objective is to achieve sufficiently long confinement times τ in
order to satisfy Lawson's criterion on the product $n\tau$. In the
laser compression approach, the objective is to achieve sufficiently
high densities, the 'confinement' time being just the natural time
scale for the plasma to expand against inertial forces.

Originally lasers had been considered only as a means of heating
plasma to high temperatures and various estimates of energy
required per pulse suggested a figure $10^8 - 10^9$ J, which was too
large to be practical. The key idea is to use the laser pulse
firstly to compress a spherical pellet of deuterium and tritium by
a factor 10^4 by an implosion scheme. The result is to reduce the
energy required down to the range $10^5 - 10^6$ J for a thermonuclear
reactor, with 1 kJ approximately the break-even level.

To see how this drastic reduction occurs, we consider the
competition between the rate of expansion of the compressed pellet
and the rate of thermonuclear burning. The former is governed by
the ordinary sound speed

$$C_s = \left(\frac{\gamma p}{\rho}\right)^{\frac{1}{2}} = \left(\frac{\gamma \kappa T}{M}\right)^{\frac{1}{2}},$$

M being the mean ion mass. The rate of thermonuclear burning at a
given temperature is proportional to the density, while C_s is
independent of density.

The time available for thermonuclear reactions to occur is $\simeq r/C_s$

r being the radius of the compressed pellet and the efficiency
η of burning is thus a function of ρr

$$\eta = \eta(\rho r).$$

ρr is an important parameter, since it also gives to a good approximation the size of the compressed region in terms of the length
scale for transport processes such as heat conduction, as well as
the mean free path for the passage of radiation, α-particles and
neutrons.

The mass μ of the compressed material is given by

$$\mu = \frac{4}{3} \pi \rho r^3 \simeq \frac{(\rho r)^3}{\rho^2}.$$

Hence, for a given efficiency η, $\mu \simeq \frac{1}{\rho^2}$ and the conclusion is that
much smaller masses are possible if high compression is feasible.

In the remaining sections an account is given of the physical
processes involved in laser-plasma interaction, the compression
phase, the generation of magnetic fields, and computer calculations.
The chapter ends with a brief mention of some interesting new fields
of research involving superdense plasma.

7-2 LASER-PLASMA INTERACTION

When a small amplitude high frequency electromagnetic wave prop-
agates through a plasma, the most important absorption process is
normally inverse bremsstrahlung in classical electron-ion collisions.
If the wave encounters a region of increasing electron density such
that the local plasma frequency approaches the wave frequency, the
wave will be almost totally reflected. When a plasma is generated
by focusing a laser beam onto a solid target, the density gradient
will usually be extremely steep, resulting in little penetration,
low absorption and only a small fraction of the incident energy
transferred to the plasma. This process becomes efficient only if
the scale length of the density gradient is larger than the product
of the electron-ion collision time with the velocity of light.

If the amplitude of the wave is increased sufficiently, the interaction with the plasma becomes strongly non-linear and several new
effects are theoretically predicted. The rate of energy absorption
by inverse bremsstrahlung reaches a maximum when the velocity of
electrons oscillating in the radiation field equals their thermal
velocity. The reason is that the collision cross section is
inversely proportional to the square of the velocity. However, in

regions where the plasma frequency approaches the wave frequency,
collective effects should become important. Parametric interactions can occur between the incident wave and plasma electron
and ion oscillations, resulting in anomalously strong absorption.
At higher intensities, other non-linear effects are also predicted
including Brillouin scattering, in which an ion acoustic wave
interacts with incident and reflected light, and Raman scattering,
in which electron plasma oscillations interact with the incident
and reflected light.

These and other processes of energy absorption are summarised
in the table below, and illustrated schematically in Fig. 7.1.

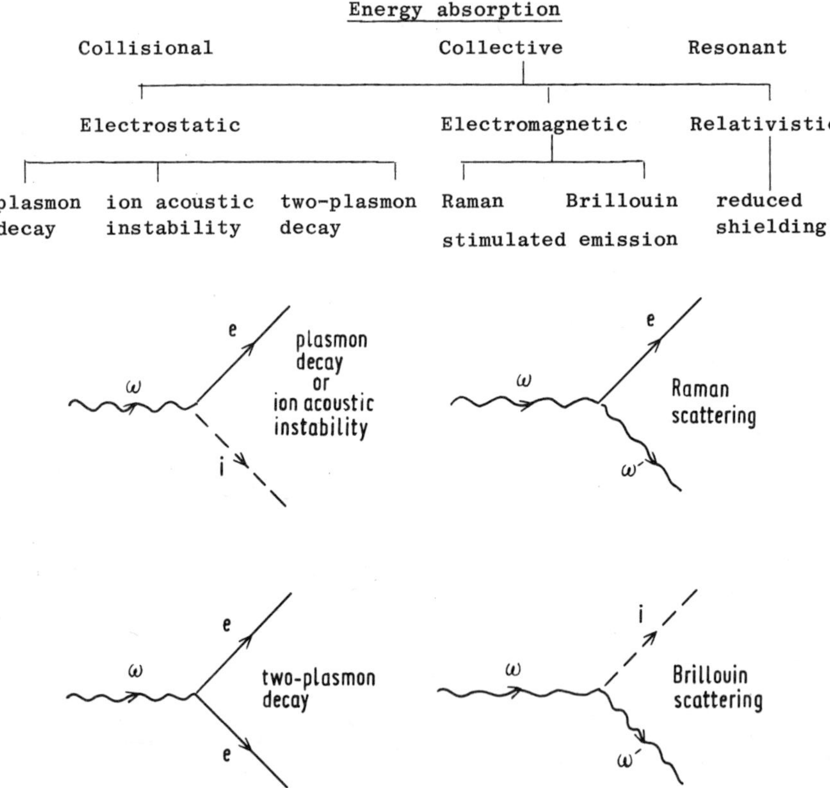

Fig. 7.1 Collective processes in energy absorption

7-3 COMPRESSION PHASE

It is important to realise the enormous pressures generated by
matter compressed to 10^4 times solid density at normal pressures.
We look first at the astrophysical situation. Hydrogen in the
solar core is believed to be about one hundred times solid density,
at a temperature 1-2 $\times$ 10^7 K and a pressure greater than 10^{11}

atmospheres. This pressure is maintained gravitationally by the
enormous overlying solar mass, approximately 10^{30} kg. In white
dwarf stars, densities of 10^8 kg m^{-3} are believed to exist, at a
pressure of 10^{15} atmospheres. In this state of matter, electrons
must be treated quantum mechanically, and in white dwarfs are
virtually Fermi degenerate. The pressure of a Fermi gas is
related to temperature by an equation of state

$$p = \frac{2}{3} n_e \, \varepsilon_F \left[\frac{3}{5} + \frac{\pi^2}{4} \left(\frac{\kappa T}{\varepsilon_F} \right)^2 + \ldots \right].$$

Here, n_e is the electron density and ε_F is the Fermi energy:

$$\varepsilon_F = \frac{h^2}{8m} \left(\frac{3}{\pi} \, n_e \right)^{\frac{2}{3}}.$$

At 10^4 times solid density, $n_e = 5 \times 10^{32}$ m^{-3} and the minimum
pressure of 10^{12} atmospheres occurs when $\kappa T \ll \varepsilon_F$, with the electrons
Fermi degenerate. This effect is essentially quantum mechanical.
In a classical system, $p = n_e \kappa T$, and the minimum pressure is zero.

In order to achieve the desired high compression, we now see that
the minimum requirement is to develop a pressure of 10^{12} atmospheres.
Our starting point is a D-T pellet with $n_e = 5 \times 10^{28}$ m^{-3} at a
temperature near absolute zero, so that the electrons are initially
Fermi degenerate. The main idea is to compress the pellet for as
long as possible without heating it appreciably, that is, we
require adiabatic compression. Notice that in this compression
ions play virtually no part, so that all the work done is concentrated on compressing the electron gas. Further, although the
electrons are nearly degenerate, the adiabatic law $p\rho^{-\gamma}$ = constant
still applies rigorously, and $\gamma = \frac{5}{3}$ as for a monatomic classical gas.

Analysis of the adiabatic compression shows that the 'piston'
must be carefully tailored as a function of time, which means that
the laser pulse itself must be very carefully shaped, with most of
the energy input in the last few picoseconds of a 100 picosecond
pulse. A simple derivation assumes spherical symmetry throughout
the compression and requires that the velocity of the imploding
surface of the pellet should never exceed the local sound speed.
If the last condition is not fulfilled shocks will develop ahead
of the 'piston' and preheat the material. Note too that adiabatic
compression can of course go as slowly as we wish. The need for a
maximum rate of compression becomes vital only when we consider the

subsequent dramatic rise in temperature when the main laser energy
is finally released into the pellet to raise the temperature to
thermonuclear levels. There is much sophistication (and heated
discussion) in this last stage, and as yet there are no absolute
experimental tests.

To go back to adiabatic compression, consider the compression of
a pellet of radius r_o, density ρ_o. The adiabatic condition is

$$p\rho^{-\gamma} = p_o\rho_o^{-\gamma}. \tag{7.1}$$

The maximum implosion velocity gives

$$\frac{dr}{dt} = - C_s = - \left(\frac{\gamma p}{\rho}\right)^{\frac{1}{2}}. \tag{7.2}$$

Conservation of matter gives

$$\rho r^3 = \rho_o r_o^3. \tag{7.3}$$

Letting C be the initial sound speed $(\gamma p_o/\rho_o)^{\frac{1}{2}}$ and incorporating
(7.3) into (7.2), we have

$$\frac{dr}{dt} = - C \frac{r_o}{r} \tag{7.4}$$

with solution

$$r^2 = r_o^2 - 2Cr_o t.$$

Hence the compression after a time t is given by

$$\alpha = \left(\frac{r_o}{r}\right)^3 = \left(1 - \frac{2Ct}{r_o}\right)^{-\frac{3}{2}}. \tag{7.5}$$

The compression follows this time-dependence until t approaches
$r_o/2C$; the energy is then ' dumped' when α reaches the required
level of compression.

We see then firstly that a pressure of 10^{12} atmospheres will be
required and that the pressure pulse has to be very carefully
shaped. The latter, though difficult, can be achieved. However
there is no question of obtaining such pressures purely from
radiation pressure of a laser beam. Laser light has been focused
to intensities exceeding 10^{21} W m^{-2}. At such intensities the
Poynting vector gives a momentum flux of 10^8 atmospheres. In
order to achieve a figure of 10^{12}, we must look for another
mechanism. In fact, we find the desired enhancement from two
different effects. The first is ablation of the pellet surface,
that is, a corona of hot plasma is generated and literally 'blasts

off' like a rocket.   The reaction drives the pellet core inwards.
The second effect is spherical convergence.   The required compression of 10^4 means a reduction by a factor of $\simeq 20$ in the
pellet radius.

In all this, we have assumed that spherical symmetry is retained
throughout the compression phase.   To achieve this, the implosion
pressure must be applied with a high degree of spherical symmetry.
One method is to split the laser beam into a series of beams
convergent on the centre of the pellet, for example, 6 beams
directed inwards along three mutually perpendicular directions.

One danger in the compression phase is the onset of Rayleigh-
Taylor instabilities, caused by the strong deceleration of plasma
as it approaches the centre of the pellet.   This acts in a way
analogous to the instability of a dense liquid resting on top of
a liquid of lower density.   Computer calculations have predicted
that such modes are unlikely to grow.   However, an interesting
development is the proposal to use hollow spheres, which for the
same total mass present a larger surface area and hence a lower
intensity requirement for the laser pulse (the total energy in the
pulse is clearly unchanged).   Here, computer predictions are much
less favourable, suggesting that the shell thickness should be at
least one-third of the pellet radius.

7-4   GENERATION OF MAGNETIC FIELDS

The simple Ohm's law assumed in §3-2

$$\underline{E} + \underline{v} \times \underline{B} = \eta \underline{j}$$

requires modification due to the Hall effect (an electric field is
set up perpendicular to both current and magnetic field) and to
finite electron density gradients.   The complete form for Ohm's law
is

$$\underline{E} + \underline{v} \times \underline{B} = \eta \underline{j} + \frac{1}{ne} \underline{j} \times \underline{B} - \frac{T}{ne} \nabla n, \qquad (7.6)$$

where n is the electron density, T the electron temperature.

It is the density gradient term which becomes crucial in the
argument.   Using Maxwell's equation

$$\frac{\partial \underline{B}}{\partial t} = - \nabla \times \underline{E} \qquad (7.7)$$

we see that any non-zero contribution to the RHS of (7.7) gives rise
to magnetic fields.   In particular,

$$\nabla \times \left(\frac{T}{ne} \nabla n \right) = \frac{1}{ne} \nabla T \times \nabla n. \qquad (7.8)$$

In (7.8) spherical symmetry ensures that ∇T and ∇n are parallel,
and no contribution arises. Any lack of spherical symmetry will
give rise potentially to very high magnetic fields. These fields
are expected to have a marked detrimental effect on the thermal
conduction in the plasma, possibly increasing any asymmetry. On
the other hand, laser energy absorption could be enhanced. The
overall conclusion is that a combination of asymmetry and poor
thermal conduction in the compression zone could prevent the attain-
ment of fusion conditions.

A tentative estimate of rise times of magnetic field from the
degree of spherical asymmetry which could be expected to occur
suggests that magnetic field of order 10^2 - 10^3 tesla could arise
spontaneously. Deliberate excitement of fields of such high
intensity would lead to very novel plasma and spectroscopic
phenomena, even though they fall short of pulsar fields.

7-5 COMPUTER CALCULATIONS

The emphasis in this section will be on the physics incorporated in
computer codes rather than on numerical algorithms. Several
authors have given very complete accounts of the numerical techniques,
and the chapter following, while not giving a specific account of
laser compression of plasma, should give sufficient indication of
the types of numerical problems encountered.

One of the most successful codes, known as LASNEX, was used in a
calculation performed on the CDC 7600 computer at Livermore Laboratory, California. LASNEX is a two-dimensional axially symmetric
finite difference code which includes detailed models of many
physical processes. Some of these are listed below.

Hydrodynamics

 Lagrangian; viscous effect including Von Neumann artificial
 viscosity; electron, ion, photon, magnetic and alpha particle
 pressures.

Laser

 Refractive transport; absorption via inverse bremsstrahlung and
 plasma instabilities; reflection at critical density.

Suprathermal electrons

 Multigroup flux-limited diffusive transport with self-consistent

electric field; non-Maxwellian electron spectra determined by
results of plasma simulation calculations for laser light absorption by plasma instabilities; inverse bremsstrahlung electron
spectrum for classical absorption.

<u>Thermal electrons and ions</u>

Flux-limited diffusion transport.

<u>Magnetic field</u>

All classical charged particle transport coefficients include
modifications due to the presence of a strong magnetic field.

<u>Radiation</u>

Multi-group flux-limited diffusive transport; Fokker-Planck
treatment of Compton scattering.

<u>Fusion</u>

Reaction rates averaged over Maxwellian velocity distribution;
the α-particle from DT reactions is transported via a one group
flux-limited diffusion model with appropriate energy deposition
into electrons and ions; one group transport of the 14 MeV
neutrons; super-dense pellet results checked against more
sophisticated thermonuclear physics codes.

<u>Material properties</u>

Opacities, pressures, specific heats, and other properties of
matter, are used which take into account nuclear, Coulomb, Fermi
degeneracy, partial ionisation, quantum mechanical, relativistic
and other significant effects.

The list, taken more or less intact from a Conference Report by
Thiessen et al.,[33] gives some idea of the scope and thoroughness of
this work. The results of these computer calculations have shown
that adequate absorption, transfer, and symmetry can be achieved in
the case of fairly short laser wavelength, such as the 1.06 μm
neodymium glass laser wavelength. Adequate laser energy absorption
is possible via inverse bremsstrahlung, occurring largely at the
critical density, near the surface of the plasma atmosphere around
the pellet where the laser optical frequency equals the electron
plasma frequency. The absorbed energy is then transferred by
electron thermal conduction to the surface of the pellet. One
point to add is that the term 'flux-limited' appears several times
in the list of physical processes incorporated in LASNEX. The
meaning of this is clear if we consider the usual formulation of the
heat flux as $\underline{q} = K \nabla T$. The temperature gradient can be so large

that the value of $\underline{q}$ calculated exceeds the maximum heat energy
which can be transported, assuming that all the electrons
travelled in the direction of the temperature gradient. In that
case the maximum electron flux is used to calculate an upper limit
to q.

7-6 CONCLUSION

Many uncertainties are still left, especially with the present lack
of experimental results. So far there is indication that a
compression ratio around 200 has been achieved. However one
disturbing feature is the detection of very high energy electrons
and ions, accompanied by hard X-ray emission. The presence of
relativistic electrons is particularly disturbing, since they can
run ahead of the compression and pre-heat the pellet core, while
the presence of ions and X-rays of energy around 100 keV suggests
violent electrostatic acceleration, possibly due to large amplitude
turbulence.

Most of this chapter has dealt with superdense plasma in the
context of fusion. It is therefore of interest to end the
discussion with a few remarks about other avenues of research opened
up by the availability of superdense plasma, and the possibility
of new technology.

<u>New fields of physics</u>

   (i)   Superdense Fermi degenerate plasma

  (ii)   Passage of relativistic electron beam

 (iii)   Emission spectra in the presence of intense magnetic fields

  (iv)   Radiative and other transport phenomena

   (v)   Nuclear synthesis.

   (i)   opens up a new world of theory and experiment, the application
of quantum statistical field theory, the possibility of superconducting properties, and so on.

  (ii)   is a problem in quantum electrodynamics and we should also
include the passage of radiation. Effects such as vacuum
polarisation and self-energy infinities are no longer
theoretical ploys, but become real.

 (iii)   is an exciting new field of atomic physics.

  (iv)   is of interest to astrophysics as also is (v).

134

<u>New technology</u>

  (i)   Intense neutron and X-ray sources.

 (ii)   Development of ultra-violet and X-ray lasers.

(iii)   Development of picosecond and femtosecond diagnostic
techniques.

Chapter 8

COMPUTATIONAL PLASMA PHYSICS

8-1   INTRODUCTION

In this closing chapter, a short account will be given of the part
currently being played by the computer both in obtaining a better
understanding of plasma phenomena and in actually simulating real
plasma experiments.

Lack of agreement between theory and experiment can be due to
various causes :

(a)   inaccuracies in calculation with an otherwise correct but
      very complex theory

(b)   invalid simplifications (geometric or otherwise) in the theory

(c)   fundamental omissions from the theory.

A parallel list could be drawn up for experimental work.

It is here that we have an important role for computational
plasma physics, namely, <u>bridging</u> the gap.   If computational plasma
physics achieved no more than this bridging between theory and
experiment, it would still have been well worthwhile.   However, so
far we are treating the computer as a logical extension of the desk
calculator.   It can also play quite a different role, which we
shall describe as <u>computer simulation</u> of plasma experiments and
phenomena.

Two examples of computer simulation will indicate the general
area of application.   In the first example, the computer simulates
real experiments using a magnetofluid model.   Many theoretical
assumptions are admittedly already built into the set of magnetofluid equations, but thereafter no further approximations are made
either in the physics or geometry.   The difficulties involved in
deriving suitable numerical schemes will be discussed in §8-2-1.

As a second example, the computer is programmed to follow in
detail the motion of a large number of ions and electrons, starting
with some given initial state, and using only Newton's laws and
Maxwell's equations of electromagnetism.   We shall discuss some of
the problems associated with particle simulation in §8-3-3.

136

Usually, a magnetofluid simulation is more able to simulate real devices in finite geometry, as in cylindrical or toroidal devices, while a particle simulation is better suited to explore particular plasma phenomena, for example charge shielding, micro-instabilities, and so on. There is, however, no hard and fast rule, and both approaches can play either a fundamental role, or a practical simulation of a real experiment, depending on the nature of the problem.

In both these areas of plasma computer simulation, diagnostics form a very important part of the work, just as in real experiments. Indeed, the same basic program can be used by different groups, interested in different experimental situations. They will then write their own diagnostic programs and "plug into" the main program.

This touches one other important aspect of computational physics research, namely, the need for compatibility between different programs, and in fact between different computers. There are many obvious examples in industry where components must all fit and be interchangeable, and there is no reason why computer programs should not follow this pattern. This then leads naturally to the require-ment of clear documentation.

Clear presentation of results is vital. It is much easier to appreciate the behaviour of, say, a Bessel function from a graph than from a set of numerical tables, though both are necessary. Thus, depending on the situation, good graphical presentation is essential.

It is a simple fact of life that computer simulation of plasma phenomena will be limited by the size and speed of the computers available. While different problems must in the end be separately assessed, we can obtain a good guide by considering a typical situation in a magnetofluid calculation. We shall suppose that reasonable resolution can be obtained if we represent any of the quantities $\underline{v}$, $\underline{B}$, ρ and T (the velocity and magnetic fields, mass density and temperature) at 64 points for each space variable. Since we are usually interested in following the time evolution of the system, we have also to make some reasonable estimate of the number of time steps needed. We take here 300 time-steps, recognising that this may be too small by an order of magnitude or more in any particular case.

Thus, we have for two-dimensional calculations $64 \times 64 \times 300 \simeq 10^6$ space-time points and the simplest fairly complete code may require $\simeq 10^3$ calculations per space-time point. A present-day medium sized computer can accommodate such a calculation ($\simeq 1\ \mu s$ per instruction) in 20 minutes. In three dimensions we would have to increase this time by another factor of 64, and storage in the central processor would also be at a premium. Such a calculation is just possible on a large computer such as the IBM 370/195. The new generation of computers should certainly render feasible such calculations, which are becoming more and more necessary for all kinds of reasons.

One of the most spectacular efforts in computational plasma physics in recent times is in the field of laser produced super-dense plasmas. A short account of the physical principles involved was given in Chapter 7.

At the end of the book a few references[34,35,36,37] are listed which give a much more detailed and comprehensive account of the topics covered in this short chapter.

Reference (34) gives a very comprehensive survey of computational plasma physics. It is of interest to mention briefly some of its contents. It deals, among other topics, with the electrostatic sheet model (Dawson) which was the fore-runner of many 2D particle codes also discussed in the book. There are chapters on the solution of the Vlasov equation by transform methods, by finite-difference techniques, by the "waterbag method", and by application of Hamilton's principle in dynamics. Finally there is a chapter on magnetofluid calculations (Roberts and Potter) and one on the solution of a Fokker-Planck equation for a mirror-confined plasma (Killeen and Marx).

We shall not deal with the latest sophistications of numerical analysis, necessary though they are to the specialist, but simply give some idea of the range of problems which can be tackled at present, and the way some of the numerical obstacles can be overcome.

8-2 MAGNETOFLUID CALCULATIONS

A model set of magnetofluid equations is given below.[38] The transport coefficients are in general complicated functions depending on both the magnitude and direction of the magnetic field $\underline{B}$ and on the temperature T. In some cases it will be seen that the

equations have been considerably modified from the ideal set of
equations introduced in Chapter 3.   These refinements are usually
essential if one is to be successful in the simulation of real
experiments.

(i)   Mass conservation (equation of continuity)

$$\frac{\partial \rho}{\partial t} + \nabla \cdot (\rho \underline{v}) = 0.$$

(ii)   Momentum conservation (force equation)

$$\frac{\partial \underline{v}}{\partial t} + (\underline{v} \cdot \nabla) \, \underline{v} = \frac{1}{\rho} (\underline{j} \times \underline{B} - \nabla p + \nabla \cdot \underline{\underline{V}}),$$

where $\underline{\underline{V}}$ is the viscous part of the stress tensor

(iii)   Energy conservation (temperature equation)

$$\frac{\partial T}{\partial t} + \underline{v} \cdot \nabla T + (\gamma - 1) T \, \nabla \cdot \underline{v}$$

$$= \frac{\gamma - 1}{nk} \left[\underline{\underline{V}} : \nabla \underline{v} + \underline{j} \cdot (\underline{E} + \underline{v} \times \underline{B}) + \nabla \cdot (\underline{\underline{K}} \cdot \nabla T) \right],$$

where $\underline{\underline{K}}$ is the coefficient of thermal conductivity

(iv)   Conservation of magnetic flux

$$\frac{\partial \underline{B}}{\partial t} = - \nabla \times \underline{E}.$$

(v)   Conservation of electric flux

$$\nabla \times \underline{B} = \mu_o \underline{j} + \frac{1}{c^2} \frac{\partial \underline{E}}{\partial t} .$$

(vi)   Generalised Ohm's law

$$\underline{E} + \underline{v} \times \underline{B} = \underline{\underline{\eta}} \cdot \underline{j} - \frac{1}{ne} \nabla p + \frac{m_e}{ne^2} \frac{\partial \underline{j}}{\partial t} ,$$

where $\underline{\underline{\eta}}$ is the resistivity

(vii)   Equation of state

$$p = nкT = \frac{\rho к T}{m_i}$$

These equations can often be considerably simplified in particular
circumstances, but in general they should form the basis for most
magnetofluid calculations.   The transport coefficients are not
given in detail here.   Further, the pressure has been taken as a
scalar quantity, but in a sufficiently large magnetic field the
plasma may exhibit strong anisotropy and the assumption of scalar
pressure will be invalid, as we discussed in §3-8.

In any simulation of a plasma device, it may also be necessary
to consider as a further refinement of the model equations the
inclusion of the effect of partial ionisation, and of impurity ions.

8-2-1 One-dimensional calculations (1D)

During the last 10-15 years, realistic calculations have been
performed in one space dimension + time (x,t) on the dynamics of
cylindrical plasma devices, especially the θ-pinch and z-pinch.
Here x stands for the radial coordinate.

One of the first serious attempts at developing a 1D code was
by Hain and Roberts[39] in 1960. They produced the first 1D
implicit Lagrangian code. (The meaning of the terms _implicit_ and
Lagrangian is discussed later.) At that time linear theta and
axial pinches were of current interest, and these were effectively
one-dimensional far enough away from the ends, at least in their
early pre-unstable regimes. In their code, radial motion was
followed in the initial implosion stage of a fully ionised theta
pinch. The code successfully simulated many features of the
experimental results. For example, it was found experimentally
that the plasma in theta pinches tends to form an annulus, with a
trapped magnetic field inside the plasma, itself contained by a
large external field. The annular plasma oscillates radially, and
good agreement was obtained between experimental results and the
code, for the frequency and damping of the oscillations.

Roberts[40] in 1963 modified the Hain code to include a neutral
component, more realistic transport processes and heat exchange
between ions and electrons. This became a powerful tool in understanding the physical processes occurring during the partially
ionised initial stage of the theta pinch.

The Hain code, and its modifications, provided therefore a
fairly general solution to one-dimensional plasma motion, and is
still much used.

We shall discuss the two main numerical problems encountered in
one-dimensional calculations and surmounted in the Hain, Roberts
code. Alfven waves are propagated with speed $C_A = (B^2/\mu_o\rho)^{\frac{1}{2}}$. One
of the main differences with fluid dynamics, so far as the solution
of practical problems is concerned, is that regions of normal
plasma density and field strength are often surrounded by regions
of high field and very low density, where C_A is much higher. The
Alfven speed can then exceed other characteristic velocities of
propagation (e.g. sound speed and rate of diffusion of plasma
across the magnetic field) by several orders of magnitude. This
imposes a severe restriction on the maximum time step which can be

used in _explicit_ calculations, leading to excessive computer time.
This restriction in its simplest form is $\Delta t < \Delta x/C_A$, and can be
interpreted in the following way. We wish to replace the continuous variables x, t and partial derivatives by discrete variables with a spacing Δx, Δt and by differences, respectively. We
may call this the 'discretisation' of the equations. A solution
of the difference equations can propagate information across the
(x,t) mesh at a maximum rate of $\frac{\Delta x}{\Delta t}$. This must be able to keep
up with any _physical_ velocity of propagation, such as a wave
velocity or diffusion velocity. If C_A is the fastest character-
istic speed, then $\frac{\Delta x}{\Delta t} > C_A$, i.e. $\Delta t < \Delta x/C_A$. In the case of a
diffusion process, suppose D is the characteristic diffusion
coefficient. We have in mind here a diffusion equation of the
form

$$\frac{\partial y}{\partial t} = D\,\frac{\partial^2 y}{\partial x^2}$$

(strictly this will be only part of an equation involving both
diffusion and wave propagation). In this case, the restriction
on Δt is given by

$$\Delta t < \frac{\Delta x^2}{D}\ .$$

The remedy for this situation, which will allow us to relax the
restriction on Δt, is to adopt an _implicit_ numerical scheme. The
difference between explicit and implicit codes is a central issue
in the solution of the partial differential equations of magnetofluid theory.

A much simpler problem, that of heat flow, (Richtmyer and
Morton)[36], illustrates very simply and clearly this difference.

Consider the unidirectional flow of heat in a large slab of
homogeneous material of length L. The temperature is taken to be
uniform on each face perpendicular to the flow of heat.

Let the temperature be u(x,t), satisfying the equation

$$a\,\frac{\partial u}{\partial t} = \frac{\partial}{\partial x}\left(K\,\frac{\partial u}{\partial x}\right),$$

where a is the heat capacity per unit volume of material and K is
the thermal conductivity, both constant. Then define $D = K/a > 0$
= diffusion coefficient.

We shall suppose that the faces of the slab at x = 0 and L are
maintained at some constant temperature, so that the boundary
conditions for all t are

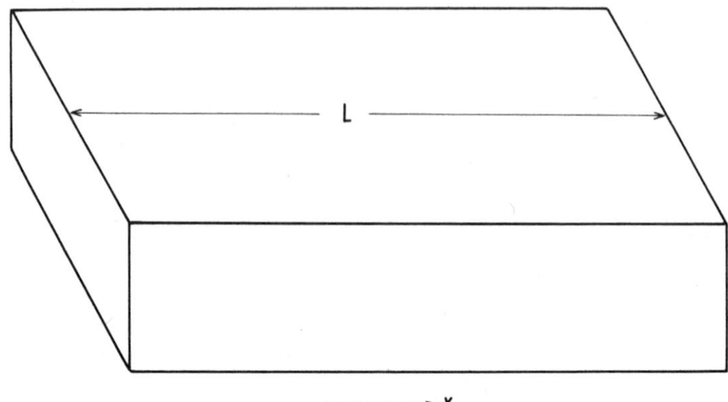

$$u(0, t) = u(L, t) = T_o$$

and for convenience we shall use a scale with $T_o = 0$. We further suppose that initially the temperature increases linearly and symmetrically from both ends, so that the initial condition is

$$u(x,0) = 2\frac{x}{L}T, \qquad 0 \leqslant x \leqslant \tfrac{1}{2}L$$

$$= 2\frac{(L-x)T}{L}, \qquad \tfrac{1}{2}L \leqslant x \leqslant L.$$

The problem is to determine the temperature distribution at subsequent times. This is, of course, a trivial problem in analysis. An exact solution can be obtained quite readily in terms of a Fourier series. We shall discuss what happens when the partial differential equation is solved numerically using finite differences as approximations to the partial derivatives.

Let Δx, Δt be increments of x, t

$\Delta x = L/J$, J some large even integer.

The (x,t) plane becomes the mesh $x = j\Delta x$, $t = n\Delta t$ with $j = 0$, $1 \ldots J$ and $n = 0, 1 \ldots$.

Denote by u_j^n the approximation to $u(j\Delta x, n\Delta t)$. The simplest <u>difference</u> equation replacing the original equation for u is

$$\frac{u_j^{n+1} - u_j^n}{\Delta t} = D\frac{u_{j+1}^n - 2u_j^n + u_{j-1}^n}{(\Delta x)^2},$$

where $j = 1, 2 \ldots J-1$ and $n = 0, 1 \ldots$

$$\text{i.e.} \quad u_j^{n+1} = u_j^n + D\frac{\Delta t}{(\Delta x)^2}\left(u_{j+1}^n - 2u_j^n + u_{j-1}^n\right).$$

Boundary conditions: $u_o^n = u_J^n = 0$, $n = 0, 1 \ldots$

Initial conditions: $u^0_j = \dfrac{2j}{J} T$, $\qquad 0 \leqslant j \leqslant \tfrac{1}{2}J$

$$= \dfrac{2(J-j)T}{J}, \qquad \tfrac{1}{2}J \leqslant j \leqslant J.$$

These equations can be used recursively to determine all u^1_j, then all u^2_j, ... for $0 < j < J$. Two sets of results are shown in Fig. 8.1, both with $J = 20$, but with slightly different Δt.

In (i), $\dfrac{D\Delta t}{(\Delta x)^2} = \dfrac{5}{11}$, while in (ii) $\dfrac{D\Delta t}{(\Delta x)^2} = \dfrac{5}{9}$.

(i)   gives perfect agreement with the exact solution, shown as a solid line.

(ii)   the agreement is never satisfactory and after only several Δt errors grow to an unacceptable level.

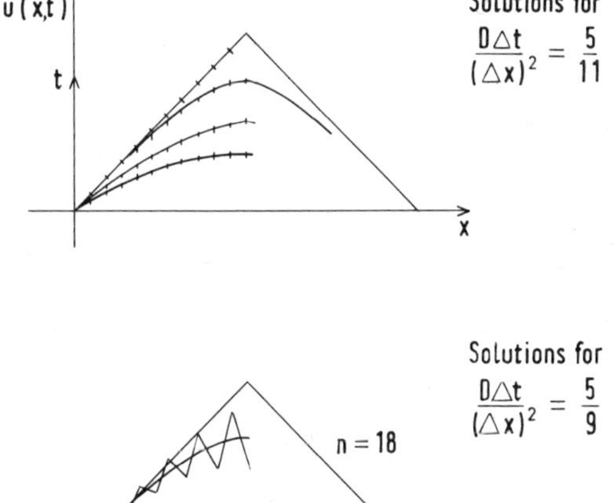

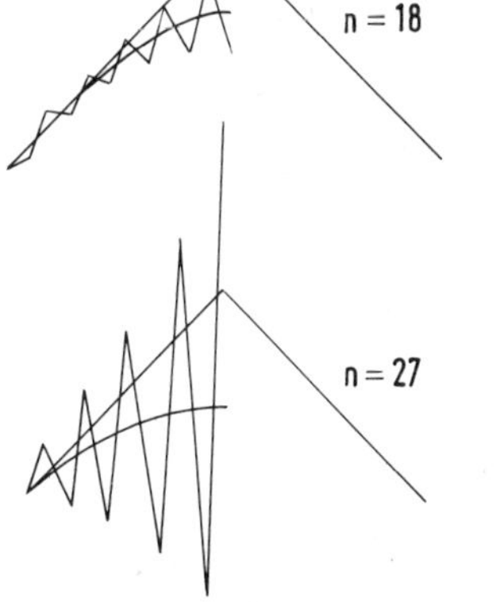

Fig. 8.1 Numerical instability in explicit solution of a diffusion equation

This phenomenon is <u>numerical</u> instability and has nothing to do with rounding-off errors as such. It is an intrinsic property of the <u>exact</u> solution of the <u>difference</u> equations. Such limitations in difference equations have been known for a long time - it is

143

interesting in passing that they were first discussed in a paper by Courant et al.[41] in 1928, the year Langmuir reported the discovery of plasma oscillations.

Note that taking smaller Δx would make matters worse. In fact, the stability criterion, which we shall derive, is

$$\frac{D \Delta t}{(\Delta x)^2} < \frac{1}{2} \, ,$$

which is satisfied by (i), but not by (ii).

When D is a function of position $D(x)$, essentially the same kind of numerical instability occurs, so that if $D(x)$ is very large in places, Δt must be correspondingly small. This is what happens in an explicit calculation, so called because the difference equations contain the "new time" (n+1) only on the left hand side, so that once we know u_j^n for all j we get each u_j^{n+1} by the single equation, given earlier.

The standard way of analysing numerical stability is like that of normal modes of oscillation. The simplest way is to substitute the expression $A \xi^n(m) \exp(imj \, \Delta x)$ for u_j^n. Here $\xi(m)$ is an amplification factor, raised to the power n. We then find

$$\xi(m) \; = \; 1 - \frac{2D \, \Delta t}{(\Delta x)^2} (1 - \cos m \Delta x).$$

Stability is assured if $|\xi| < 1$, all m. For our problem the criterion is $\dfrac{2D \, \Delta t}{(\Delta x)^2} < 1$.

Implicit difference schemes

In the original difference scheme (or algorithm) above we chose to replace $\dfrac{\partial^2 u}{\partial x^2}$ by $\dfrac{(u_{j+1} - 2u_j + u_{j-1})_{t=n \Delta t}}{(\Delta x)^2}$, i.e. to evaluate $\dfrac{\partial^2 u}{\partial x^2}$ at time-level n.

There is no reason in principle for choosing this time and not (n+1), or a combination of the two. Suppose then we choose the algorithm

$$\frac{u_j^{n+1} - u_j^n}{\Delta t} \; = \; \frac{D}{(\Delta x)^2} \left[u_{j+1}^{n+1} - 2u_j^{n+1} + u_{j-1}^{n+1} \right] .$$

It is obvious why this was not our first choice, for it is no longer a simple matter to update from n to (n+1), i.e. knowing all u_j^n does not allow us easily to get u_j^{n+1}. We now have a set of simultaneous equations to solve, and the system of equations is called implicit.

However, the interesting aspect of this scheme, though more complicated, is revealed by carrying out the same stability analysis as before.   We find now

$$\xi(m) \;=\; \left[1 + \frac{2D\Delta t}{(\Delta x)^2} \, (1 - \cos m\Delta x) \right]^{-1} ,$$

which shows that $\xi(m) \leqslant 1$ for all m, regardless of the value of $\frac{D\Delta t}{(\Delta x)^2}$.   So this scheme is always stable;   the only restriction on Δx and Δt is to ensure accuracy.   Clearly Δx and Δt must be sufficiently small that our approximations for $\frac{\partial u}{\partial t}$ and $\frac{\partial^2 u}{\partial x^2}$ are accurate.   We can interpret this result in terms of the rate of propagation of information.   In an explicit scheme, this was $\frac{\Delta x}{\Delta t}$. In the implicit scheme above it is effectively infinite.

One last point:

$$\text{Let} \quad (\delta^2 u)^n_j \;=\; u^n_{j+1} - 2u^n_j + u^n_{j-1} .$$

Then consider the more general scheme

$$\frac{u^{n+1}_j - u^n_j}{\Delta t} \;=\; \frac{D}{(\Delta x)^2} \left[\theta \, (\delta^2 u)^{n+1}_j + (1 - \theta) \, (\delta^2 u)^n_j \right].$$

θ is a real constant, $0 \leqslant \theta \leqslant 1$.

$\theta = 0$ is the explicit scheme.

$\theta = 1$ is the implicit scheme just discussed.

It is easy to show that the amplification factor satisfies the stability criterion in general if

$$\frac{2D\Delta t}{(\Delta x)^2} \;\leqslant\; \frac{1}{1-2\theta} \;, \quad \text{when } 0 \leqslant \theta \leqslant \tfrac{1}{2}$$

and no restriction, when $\tfrac{1}{2} \leqslant \theta \leqslant 1$.

The special case $\theta = \tfrac{1}{2}$, which is stable for all Δx, Δt is sometimes referred to as <u>time-centred</u>.   It is possible to extend these ideas to the magnetofluid, and any other, partial differential equations, in plasma physics and elsewhere, and successful numerical schemes have been developed.   One further complication is that the magnetofluid equations are non-linear (suppose for example that D in the thermal diffusion problem had been a function of temperature D(u).)   Then some sensible iterative scheme must also be devised.

Next, a few comments about advective terms.   A major problem arises in magnetofluid calculations from the dominance of the advective term $\underline{v}.\nabla f$ occurring in the equations, $\left(\frac{df}{dt} \equiv \frac{\partial f}{\partial t} + \underline{v}.\nabla f \right)$. This represents transport of f (density, magnetic field, temperature, velocity) from point to point just by the bulk motion of fluid, leaving their values unchanged.   Superimposed on this advection

145

are other kinematic effects of bulk motion, such as adiabatic compression, rotation and diffusion. In the _Eulerian_ form, the independent space variables refer to co-ordinates fixed in space through which the fluid is moving. The flow is described in terms of a time-dependent velocity field which is the solution of the hydrodynamic equations for some given initial and boundary conditions.

The _Lagrangian_ form adopts a co-ordinate system fixed in the fluid, and undergoes all the distortions of the fluid so that fluid elements are permanently identified by their Lagrangian variables, while their locations in space constitute the solution of the problem. In one dimension, the Lagrangian method has proved very successful. Essentially, the magnetofluid equations are transformed with the advective terms absent. They are then solved by some suitable numerical difference scheme.

The danger inherent in advective terms present in the Eulerian formalism can be illustrated quite simply. Consider the equation

$$\frac{\partial f}{\partial t} + v \frac{\partial f}{\partial x} = 0,$$

which is the simplest form of an advective equation. Using the same notation as before, a straightforward differencing of the equation gives

$$\frac{f_j^{n+1} - f_j^n}{\Delta t} = -v \frac{f_{j+1}^n - f_{j-1}^n}{2\Delta x},$$

i.e.

$$f_j^{n+1} = f_j^n - \frac{v\Delta t}{2\Delta x}\left(f_{j+1}^n - f_{j-1}^n\right).$$

Setting $f_j^n = A\,\xi^n(m)\,\exp(imj\Delta x)$ gives

$$\xi(m) = 1 - v \frac{\Delta t}{\Delta x} i \sin m\Delta x,$$

so

$$|\xi| = \left(1 + \left(\frac{v\Delta t}{\Delta x}\right)^2 \sin^2 m\Delta x\right)^{\frac{1}{2}} \geq 1$$

and difference equation is always unstable.

There are several ways to eliminate this, for example,

(i) Replace $\frac{\partial f}{\partial x}$ by $\dfrac{f_{j+1}^n - f_j^n}{\Delta x}$, $v < 0$

and by $\dfrac{f_j^n - f_{j-1}^n}{\Delta x}$, $v > 0$.

This is termed forward and backward differencing, respectively, and can be shown to be equivalent to replacing $v \frac{\partial f}{\partial x}$ by

$$\left(v \frac{\partial f}{\partial x} - \frac{|v|}{2} \Delta x \frac{\partial^2 f}{\partial x^2} \right) ,$$

so that we are in essence solving the equation

$$\frac{\partial f}{\partial t} + v \frac{\partial f}{\partial x} = |v| \frac{\Delta x}{2} \frac{\partial^2 f}{\partial x^2} ,$$

and RHS corresponds to diffusion, a damping and stabilising process.   The scheme is valid if $v \frac{\Delta t}{\Delta x} < 1$.

(ii)   A general prescription to "centre" all differences.   Thus, for the equation above we can use

$$\frac{f_j^{n+1} - f_j^{n-1}}{\Delta t} = -v \frac{f_{j+1}^n - f_{j-1}^n}{\Delta x} .$$

(i) gives a stable scheme but may lead to a masking of physical
     effects by too severe a <u>numerical</u> diffusion.

(ii) eliminates diffusion, but numerical <u>dispersion</u> remains, i.e.
     modes of short wavelengths travel with a velocity which is
     different from the correct speed v.

All this can of course be avoided altogether by a <u>Lagrangian</u> scheme.   The discussion above applies only to an Eulerian fixed mesh scheme.

8-2-2   Two-dimensional calculations (2D)

Calculations involving two space co-ordinates are very important
if one is to describe accurately effects of magnetic field
curvature and plasmas with highly anisotropic transport coefficients.
In thermonuclear research these effects are of vital importance in
toroidal systems.

Numerical difficulties are much more formidable than in one-
dimensional calculations.

(a) In principle, the time-step limitation for an explicit cal-
    culation can again be avoided by an implicit code, but only
    at the expense of very sophisticated programming.   One
    simple consequence is that while the time-step may now be
    chosen with only accuracy to be considered, the <u>effective</u>
    time taken by the computer may be much longer.   An extra
    factor is that the size of the arrays of variables which must
    be stored at each time-step is greatly increased by the

implicit nature of the code, since at each time all the
space dependence of the variables is used simultaneously and
no overwriting is possible.   Thus a large amount of space in
the computer's central processor unit (CPU) is required,
unless complex buffering procedures are adopted.   This again
increases the real time of the calculations.

(b) Advective terms can again lead to anomalous numerical diffusion.
Now a moving Lagrangian mesh is much more difficult to implement,
since motion is much more complicated in 2D.

(c) There is considerable anisotropy in plasma transport coefficients
parallel and perpendicular to the local magnetic field.   The
electron thermal conductivity can be several orders of magnitude
greater along the magnetic field than across it in a typical
hot magnetized plasma.   This indicates that field lines should
be used as co-ordinates, though their complex topology is a
deterrent.

There are two main approaches to magnetofluid computer calculations.   In the first approach we try to simulate a real
experiment as accurately as possible, so that the magnetofluid
equations, together with boundary and initial conditions, have to
match closely the physics and geometry of the device.   In the
second approach we choose the simplest geometry, equations and
boundary conditions that will lead to interesting results, and
then explore the consequences.

We shall describe only an example of the first kind, citing the
very successful code by Roberts and Potter[34] to simulate the
plasma focus experiment.

The plasma focus is a device (see Fig. 8.2) with axial
symmetry consisting of an inner and outer coaxial cylindrical
electrode between which an annular plasma is created.   Under the
effect of the $\underline{j} \times \underline{B}$ force, the plasma runs down axially until it
reaches the end of the shorter inner electrode.   Rapid collapse
and compression of the plasma constitutes the plasma focus.

All interesting phenomena occur in the (r, z) plane, while the
magnetic field is purely azimuthal B_θ .   All transport coefficients
are therefore isotropic and the vacuum field has simple $\frac{1}{r}$ dependence.   The physical processes are supersonic, so that an explicit
code is sufficient, consisting mainly of a shock which collapses
onto a small region near the axis (the focus) followed by axial

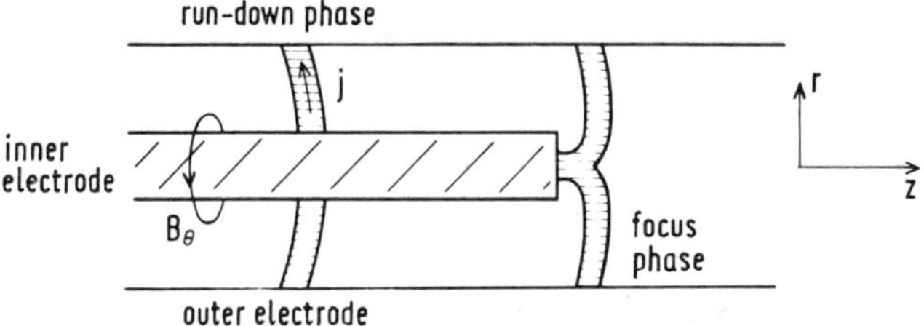

Fig. 8.2   The plasma focus

flow.   An Eulerian mesh is used, switching to a local much finer mesh near the axis when the plasma focus is about to occur.   Thus, initially Δr and Δz are taken to be 0.15 cm, while in the collapse phase these are reduced to 0.02 cm.   There are several noteworthy points concerning the physical model, and the running of the code. These bring out very clearly how important it is, in simulating a real experiment, that we should get the physics right, otherwise nothing else can be.

1. The plasma is represented by a two-fluid model, with equations of continuity, radial and axial momentum, B_θ field, electron energy and total energy.

2. A generalised Ohm's law is used, including the Hall effect and the electron pressure gradient.

3. Transport coefficients take account of cyclotron orbit effects, since $\Omega_e \tau_{ei} \gg 1$ and $\Omega_i \tau_{ii} \simeq 1$, Ω_e and Ω_i being the electron and ion gyrofrequencies, τ_{ei} and τ_{ii} the electron-ion and ion-ion collision times.   However, the pressure is still taken to be scalar, since only the (r, z) plane is involved.

4. There is a problem of ion-electron equipartition if the equipartition time $\tau < \Delta t$ used in the calculation.   This is solved by treating the equipartition term implicitly when the difficulty arises.

5. During the run-down and collapse phases, hyperbolic terms dominate (i.e. Alfven waves and sound waves), while long-time effects are controlled by the parabolic or diffusive terms.

   Success of a simulation code may be assessed somewhat arbitrarily, depending on one's initial expectations.   Nevertheless, one can claim that the plasma focus code demonstrated that a complex plasma experiment can be successfully simulated on a computer.   For it agreed closely with experiment in the following details :

1. The dynamics of the run-down and collapse.

2. The heating mechanism in the dense pinch.

3. The long-lifetime of the plasma focus. (Linear instabilities
   are expected to break up the focus in $\simeq$ 15 ns compared with
   the observed lifetime $\simeq$ 200 ns. However, the code included
   ion viscous damping of short wavelength instabilities, as
   well as non-linear stabilisation and repinch of the m = 0 mode.
   These effects appear to correlate with experimental observations.)

8-3 PLASMA KINETIC THEORY

Plasma phenomena at a microscopic level are described by an
appropriate kinetic equation. Useful information can be obtained
from the Fokker-Planck equation if relaxation to thermal equilibrium
is being investigated. As a typical example, we may mention
experiments in thermonuclear research which employ the injection of
energetic neutral atoms. A plasma is formed of initially hot ions
and cold electrons, and it is of interest to know the velocity
distribution functions of ions and electrons as a function of time,
during the build-up of the plasma (cf. reference 35).

In phenomena such as high frequency waves, Landau damping,
microinstabilities, and other processes with characteristic times
of order a plasma period, one usually assumes that particle
correlations can be neglected. One then obtains the Vlasov
equation, in which particles interact via their self-consistent
electric and magnetic fields, large-angle deflections caused by
close encounters being very rare and thus having negligible effect
at this level of description.

Some more detailed discussion of the methods of solving these
kinetic equations is given below.

8-3-1 Fokker-Planck equation

A model equation for the Fokker-Planck equation is

$$\frac{\partial f}{\partial t} + \underline{v} \cdot \nabla f + \frac{e}{m}\,(\underline{E} + \underline{v} \times \underline{B}) \cdot \frac{\partial f}{\partial \underline{v}} = \underline{D}_1\,(\underline{v}) \cdot \frac{\partial f}{\partial \underline{v}} + \underline{\underline{D}}_2\,(\underline{v}) : \frac{\partial^2 f}{\partial \underline{v} \partial \underline{v}} \ ,$$

where $\underline{D}_1$ and $\underline{\underline{D}}_2$ are, respectively, frictional and velocity
diffusion coefficients determined, in a plasma, by considering
the Coulomb long-range interaction between particles, cut off at
the Debye length separation to take account of collective effects.

There are separate equations for ions and electrons. Maxwell's

equations are coupled to the kinetic equations through source
terms

$$q = e \int (f_i - f_e) \, d\underline{v} \quad ; \quad \underline{j} = e \int (f_i - f_e) \, \underline{v} d\underline{v}.$$

Calculations have been limited to systems which are spatially
uniform and to two velocity dimensions plus time.   In such problems involving relaxation processes, it is essential to use an
implicit scheme, with no restriction on time-step for numerical
stability, for during the course of the calculation the time-step
has usually to be continually increased, in order to make reasonable
progress towards equilibrium.   For example, in the type of problem
referred to in the introduction above, the electron temperature is
initially very low and the energy transfer rate from the hot ions
is high, but decreases gradually as the electron temperature
increases.

8-3-2   Vlasov equation (1D)

We must here draw a distinction between the methods employed for
solving one-dimensional and two-dimensional problems.   In the
former case, which we discuss in this section, several methods are
available.   Three are cited below:

(a) Direct solution of the integro-differential equations, non-
    linear coupled equations in one space variable, one velocity
    variable, and the time, with the electron plasma frequency
    larger than the corresponding ion frequency by a factor
    $(m_i/m_e)^{\frac{1}{2}}$.   For a magnetised plasma, there is also a character-
    istic gyrofrequency in ratio m_i/m_e.   For the non-magnetised
    plasma, and neglect of self-consistent magnetic fields, the
    system of equations to be solved is

$$\frac{\partial f_i}{\partial t} + v \frac{\partial f_i}{\partial x} + \frac{eE}{m_i} \frac{\partial f_i}{\partial v} = 0$$

$$\frac{\partial f_e}{\partial t} + v \frac{\partial f_e}{\partial x} - \frac{eE}{m_e} \frac{\partial f_e}{\partial v} = 0$$

$$\varepsilon_o \frac{dE}{dx} = e \int_{-\infty}^{\infty} (f_i - f_e) \, dv.$$

(b) Sheet model

    Consider

$$\frac{\partial f}{\partial t} + v \frac{\partial f}{\partial x} + F \frac{\partial f}{\partial v} = 0.$$

We note that f remains constant along a trajectory which
satisfies the equations

$$\frac{dx}{dt} = v \; ; \quad \frac{dv}{dt} = F.$$

This "method of characteristics" is readily extended to the
Vlasov system given above in (a).   Computer programs have
been extensively written during the past decade for problems
of varying complexity.   Initially, the motion of several tens
of charged sheets was followed.   Nowadays, large computer
"experiments" can be set up in which many thousands of sheets
are followed.

Though much information about the plasma can be obtained by
this method, a number of difficulties arise, such as:

  (i) the crossing of sheets;

 (ii) the mass-ratio of the ions and electrons is usually
taken to be much less than the true value, say 10:1;

(iii) the fact that a system of perhaps 10^{20} particles is
being replaced by 10^{4}.

(c) Waterbag

A system of particles obeying the Vlasov equation behaves in
<u>phase space</u> (x, v) as an incompressible fluid.   Thus, following
particle motion the distribution function f remains constant.
In the case of a plasma both ions and electrons separately
have this property.   The interaction between them, through
their self-consistent fields, modifies the ion and electron
motion, but preserves the property of incompressibility in
phase space.

In the Waterbag model and its refinements, we invoke this
principle.   A distribution function is represented by a step
function or set of step functions.   The motion of the boundaries of a region in phase space of constant f is then computed
as a function of time by solving the equations of motion for a
finite set of boundary points.   While considerable twisting
and distortion of the boundaries may occur we know that
particle trajectories cannot cross out of or into the region of
constant f.   Hence the name Waterbag.   A simple example is
illustrated in Fig. 8.3.

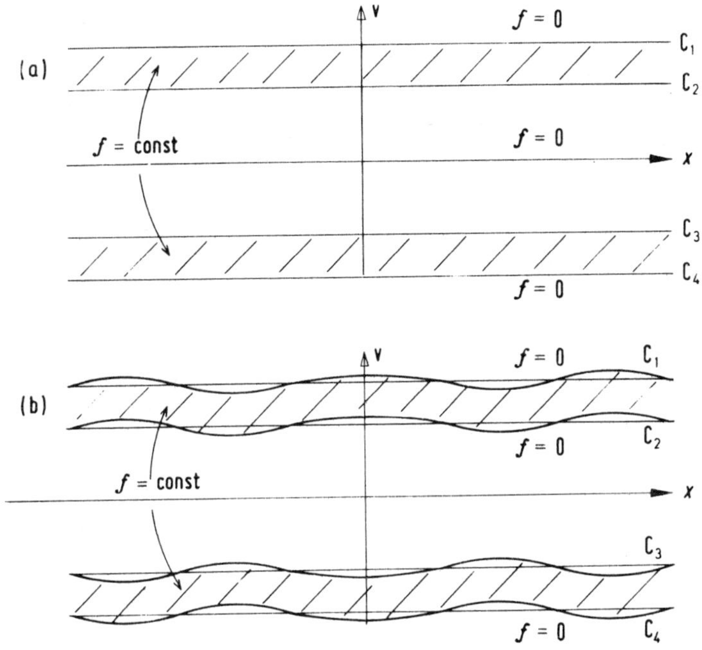

Fig. 8.3 Waterbag simulation of the two-stream instability.
(a) initial configuration   (b) development of instability

8-3-3   Vlasov equation (2D)

A straightforward extension to two dimensions is possible (i.e.
four-dimensional phase space plus time).   While direct solution of
the resulting system of equations, using finite difference schemes,
is just feasible, most of the work done so far has been to extend
the sheet model to a "rod" model.   An important feature of such
computer programs is the incorporation of a method of solving
Poisson's equation by fast Fourier transform techniques.   While
there is some degree of flexibility in such codes, the plasma is
usually bounded by a square, or at best a rectangle, with doubly-
periodic boundary conditions (sometimes replaced by perfectly
conducting walls).   An early version was the GALAXY code developed
at the Culham Laboratory several years ago by Boris and Roberts.[42]

It is instructive to follow the procedures adopted in a code of
this type by Hockney.[43] We store the phase-space co-ordinates of
all the ions and electrons at time t, $(x \ y \ v_x \ v_y)_t$, and advance
them stepwise in time.   The charge distribution and electrostatic

153

potential are computed on a finite mesh, thus:

(i) From all the known particle positions, adding for ions and subtracting for electrons, we build up the charge distribution

$$q(x, y).$$

(ii) Solve Poisson's equation for the electrostatic potential

$$\varepsilon_o \nabla^2 \phi(x, y) = -q(x,y).$$

(iii) Calculate the electric field at each grid point

$$\underline{E}(x, y) = -\nabla\phi(x, y).$$

(iv) Accelerate each particle for a time Δt

$$\Delta\underline{v} = \frac{e}{m}\underline{E}\,\Delta t \qquad \Delta\underline{x} = \underline{v}\,\Delta t,$$

giving finally $(x + \Delta x, y + \Delta y, v_x + \Delta v_x, v_y + \Delta v_y) = (x\ y\ v_x\ v_y)_{t + \Delta t}$.

The process can then be repeated, as shown below.

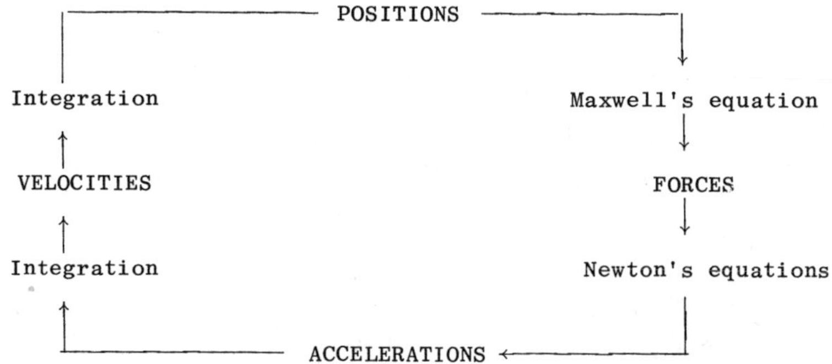

Several comments should be made.

The method of calculating the charge density is important. Although initially all particles may be located at a grid point, subsequent positions need not be. Two main methods of procedure are possible, known as the nearest grid point (NGP) and the cloud in cell (CIC). Briefly, in NGP the whole charge of a particle whose co-ordinates (x, y) lie anywhere within a cell of the mesh is associated with the grid point (at the centre of the cell). In CIC, the particle location (x, y) is considered as the centre of a uniform cloud of charge of the same size and shape as the cell. The amount of charge is then apportioned to the four surrounding grid points according to the amount of overlap of the charge cloud in each appropriate cell. Fig. 8.4 shows the differences between NGP and CIC, and the law of force F(r) which results in both approximations.

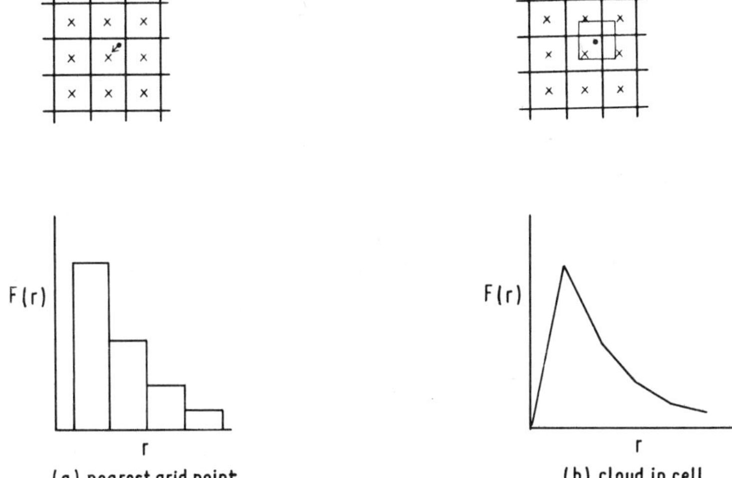

Fig. 8.4   Calculation of charge density

Though theoretically more desirable, CIC has the disadvantage
of any complex computational procedure in requiring more instructions
to implement it.   In fact, it requires approximately five times as
many as NGP.   Thus, although for the same number of particles, NGP
leads to more "numerical noise", it allows five times as many
particles to be used, which largely compensates.   The choice of
the method thus depends on other factors more closely connected to
the particular problem being considered.

Step (ii), the Poisson solver, can lead to excessive computer
time unless the fast Fourier transform techniques mentioned earlier
are employed.   These are not appropriate in any geometry more
complex than cartesian, so that any attempt to apply these methods
to toroidal systems will require the fastest and largest available
computer.

Only a self-consistent electrostatic potential has been calculated
here.   To include a corresponding self-consistent magnetic field
is feasible, requiring a vector equation for the vector potential
with the current density as source.

Step (iv) is sometimes termed the "particle pusher".   It can be
modified to take account of an externally imposed constant magnetic
field perpendicular to the plasma motion.   Another variant is the
introduction of guiding centre equations, should this be an
appropriate approximation.

155

REFERENCES

CHAPTER 1

1. cf. J.A. Ratcliffe. 1959. The Magneto-Ionic Theory (Cambridge University Press).

2. I. Langmuir. 1929. Phys. Rev. $\underline{33}$, 195.

3. J.D. Lawson. 1957. Proc. Phys. Soc. B$\underline{70}$, 6.

CHAPTER 2

4. A.J. Lichtenberg. 1969. Phase-Space Dynamics of Particles (Wiley).

5. A.A. Galeev and R.Z. Sagdeev. 1968. Sov. Phys. - JETP $\underline{26}$, 233.

CHAPTER 3

6. N.M. Rosenbluth and C.L. Longmire. 1957. Ann. Phys., $\underline{1}$, 120.

7. D.M. Meade. 1966. Phys. Rev. Letters, $\underline{17}$, 677.

8. I.B. Bernstein, E.A. Frieman, M.D. Kruskal and R.M. Kulsrud. 1958. Proc. Roy. Soc. $\underline{A244}$, 17.

9. B.R. Suydam. 1958. Proc. 2nd Geneva Conference on Peaceful Uses of Atomic Energy (Columbia University Press).

10. W.A. Newcomb. 1960. Ann. Phys. $\underline{10}$, 232.

11. G.F. Chew, M.L. Goldberger and F.E. Low. 1956. Proc. Roy. Soc. $\underline{A236}$, 112.

12. R.J. Tayler. 1970. Chapter 3 of Physics of Hot Plasmas (see Bibliography).

CHAPTER 4

13. L.D. Landau. 1946. Sov. Phys. - JETP, $\underline{10}$, 25.

14. J.H. Malmberg and C.B. Wharton. 1964. Phys. Rev. Letters $\underline{6}$, 184; 1967. Phys. Rev. Letters $\underline{19}$, 775.

15. T.H. Stix. 1962. The Theory of Plasma Waves (McGraw-Hill).

16. B.D. Fried and S.D. Conte. 1961. The Plasma Dispersion Function (Academic Press).

17. O. Buneman. 1959. Phys. Rev. $\underline{115}$, 503.

18. I.B. Bernstein. 1958. Phys. Rev. $\underline{109}$, 10.

19. W.B. Thompson. 1970. Chapter 1 of Physics of Hot Plasmas
     (see Bibliography).

20. N.N. Bogoliubov. 1946. Problems of a Dynamical Theory in
     Statistical Physics. Tr. in Statistical Physics, ed.
     J. de Boer and G.E. Uhlenbeck, 1962 (North-Holland).

21. E.A. Frieman. 1963. J. Math. Phys. $\underline{4}$, 410.

22. G. Sandri. 1963. Ann. Phys. $\underline{24}$, 332, 380.

23. N.N. Bogoliubov and N.M. Krylov. 1947. Introduction to
     Non-Linear Mechanics. Tr. by S. Lefshetz (Princeton
     University Press).

24. R. Balescu. 1960. Phys. Fluids $\underline{3}$, 52.

25. A. Lenard. 1960. Ann. Phys. $\underline{10}$, 390.

26. R.L. Guernsey. 1962. Phys. Fluids $\underline{5}$, 322.

CHAPTER 5

27. D. Bohm. 1949. Characteristics of Electrical Discharges in
     Magnetic Fields. Ed. A. Guthrie and R.K. Wakerling
     (McGraw-Hill).

28. W.H. Bennett. 1934. Phys. Rev. $\underline{45}$, 890.

29. M.S. Ioffe. 1965. In Plasma Physics (IAEA, Vienna).

30. J.B. Taylor. 1974. Proc. 5th IAEA Conf. on Fusion and
     Plasma Physics, Tokyo. Phys. Rev. Letters $\underline{33}$, 1139.

CHAPTER 6

31. P.A. Sturrock (Ed.). 1967. Proc. International School of
     Phys., Course XXXIX. Plasma Astrophysics (Academic Press).

32. D.G. Wentzel and D.A. Tidman (Ed.). 1969. Plasma
     Instabilities in Astrophysics (Gordon and Breach).

CHAPTER 7

33. A. Thiessen, G. Zimmerman, T. Weaver, J. Emmett, J. Nuckolls
     and L. Wood. 1973. Proc. 6th Euro. Conf. on Controlled
     Fusion and Plasma Physics, Moscow.

CHAPTER 8

34. Methods in Computational Physics. Vol. 9 - Plasma Physics.
     1970. (Academic Press).

35. J. Killeen. 1970. Chapter 5 of Physics of Hot Plasmas
     (see Bibliography).

36. R.D. Richtmyer and K.W. Morton. 1967. Difference Methods for Initial Value Problems. 2nd edition (Interscience).

37. D.E. Potter. 1974. Computational Physics (Wiley).

38. W.J. Sharp. 1972. Ph.D. Thesis, Glasgow University.

39. K. Hain, G. Hain, K.V. Roberts, S.J. Roberts and W. Koppendorfer. 1960. Z. Naturforsch. 15a, 1039.

40. K.V. Roberts. 1963. J. Nucl. Energy, Pt.C, 5, 365.

41. R. Courant, K.O. Friedrichs and H. Lewy. 1928. Math. Ann. 100, 32. (In German)

42. J.P. Boris and K.V. Roberts. 1969. Proc. Computational Physics Conf., Culham Laboratory (H.M. Stationery Office, London).

43. R.W. Hockney. 1969. Proc. Computational Physics Conf., Culham Laboratory (H.M. Stationery Office, London).

BIBLIOGRAPHY

J.M. Boyd and J.J. Sanderson. 1969. *Plasma Dynamics* (Nelson).

B.E. Keen (Ed.). 1974. *Plasma Physics*. Conf. Series No. 20, Institute of Physics, London.

B.J. Rye and J.C. Taylor (Ed.). 1970. *Physics of Hot Plasmas* (Oliver and Boyd).

W.B. Thompson. 1964. *An Introduction to Plasma Physics* 2nd Edition (Pergamon).

T.Y. Wu. 1966. *Kinetic Equations of Gases and Plasmas* (Addison Wesley).

INDEX